I0044327

# Modern VLSI Design

# Modern VLSI Design

Charlotte Stedman

Larsen & Keller
www.larsen-keller.com

Modern VLSI Design
Charlotte Stedman
ISBN: 978-1-64172-082-3 (Hardback)


© 2019 Larsen & Keller



⊟ Larsen & Keller

Published by Larsen and Keller Education,
5 Penn Plaza,
19th Floor,
New York, NY 10001, USA


**Cataloging-in-Publication Data**

Modern VLSI design / Charlotte Stedman.
    p. cm.
Includes bibliographical references and index.
ISBN 978-1-64172-082-3
1. Digital integrated circuits--Computer-aided design. 2. Integrated circuits--Very large scale integration.
3. Logic circuits--Computer-aided design. I. Stedman, Charlotte.
TK7874.75 .M63 2019
621.395--dc23

This book contains information obtained from authentic and highly regarded sources. All chapters are published with permission under the Creative Commons Attribution Share Alike License or equivalent. A wide variety of references are listed. Permissions and sources are indicated; for detailed attributions, please refer to the permissions page. Reasonable efforts have been made to publish reliable data and information, but the authors, editors and publisher cannot assume any responsibility for the validity of all materials or the consequences of their use.


Trademark Notice: All trademarks used herein are the property of their respective owners. The use of any trademark in this text does not vest in the author or publisher any trademark ownership rights in such trademarks, nor does the use of such trademarks imply any affiliation with or endorsement of this book by such owners.

For more information regarding Larsen and Keller Education and its products, please visit the publisher's website www.larsen-keller.com

# Table of Contents

# Preface

VLSI or very-large-scale integration is a process of designing an integrated circuit (IC) by combining a large number of transistors or devices into a single chip. The microprocessor is a common example of a VLSI device. Before the advent of VLSI design, ICs performed a limited set of functions. Modern designs employ extensive automated logic synthesis and design automation to lay out the transistors. This enables higher levels of complexity in logic functionality. High-performance logic blocks, such as SRAM cell or static random-access memory cell are manually designed to ensure maximum efficiency. This book elucidates the concepts and innovative models around prospective developments in the field of VLSI design in the modern scenario. Some of the diverse topics covered in this book address the varied aspects of VLSI systems. It aims to serve as a resource guide for students and experts alike and contribute to the growth of the discipline.

Given below is the chapter wise description of the book:

**Chapter 1**, VLSI or very-large-scale integration refers to the method of creation of an integrated circuit (IC) through the combination of a large number of transistors and devices into a single chip. This chapter has been carefully written to provide an introduction to VLSI. It includes varied aspects of integrated circuits, their design, etc. **Chapter 2**, In electronics, a logic gate is a device implementing a Boolean function or performing a logical function on a set of binary inputs and performing a single output. These are constructed using diodes or transistors. This chapter closely examines the fundamentals of logic gates and logic synthesis. It includes vital topics related to combinatorial logic, CMOS, logic synthesis, VLSI design, interconnect delay, combinational network delay, etc. **Chapter 3**, The field of electronics has undergone rapid developments in the past decades, which has resulted in innovation of electronic devices such as sequential machines. This chapter discusses the fundamentals of power optimization, clocking, flip flops, sequential systems and clock generators. **Chapter 4**, The floorplan of an integrated circuit is a schematic representation of the arrangement of functional elements. This chapter includes vital topics related to wire bonding, line driver and floor plan, which are crucial for an extensive understanding of floorplanning. **Chapter 5**, An understanding of VLSI architecture requires comprehension of the fundamentals of hardware description language, register-transfer level, pipeline in computing and globally asynchronous locally synchronous architecture. This chapter covers all these essential aspects for a thorough understanding.

Indeed, my job was extremely crucial and challenging as I had to ensure that every chapter is informative and structured in a student-friendly manner. I am thankful for the support provided by my family and colleagues during the completion of this book.

**Charlotte Stedman**

# Introduction to VLSI

VLSI or very-large-scale integration refers to the method of creation of an integrated circuit (IC) through the combination of a large number of transistors and devices into a single chip. This chapter has been carefully written to provide an introduction to VLSI. It includes varied aspects of integrated circuits, their design, etc.

## Integrated Circuit

Every electronic appliance we use in our day-to-day life, such as mobile phones, laptops, refrigerators, computers, televisions and all other electrical and electronic devices are manufactured with some simple or complex circuits. Electronic circuits are realized using multiple electrical and electronic components connected with each other by connecting wires or conducting wires for the flow of electric current through the multiple components of the circuit, such as resistors, capacitors, inductors, diodes, transistors, and so on.

Circuits can be classified into different types based on different criteria, such as, based on connections: series circuits and parallel circuits; based on the size and manufacturing process of circuit: integrated circuits and discrete circuits; and, based on signal used in circuit: analog circuits and digital circuits.

Integrated circuit or IC or microchip or chip is a microscopic electronic circuit array formed by the fabrication of various electrical and electronic components (resistors, capacitors, transistors, and so on) on a semiconductor material (silicon) wafer, which can perform operations similar to the large discrete electronic circuits made of discrete electronic components.

Integrated Circuits

As all these arrays of components, microscopic circuits and semiconductor wafer material base are integrated together to form a single chip, hence, it is called as integrated circuit or integrated chip or microchip.

Electronic circuits are developed using individual or discrete electronic components with different sizes, such that the cost and size of these discrete circuits increase with the number of components used in the circuit. To conquer this negative aspect, the integrated circuit technology was developed – Jack Kilby of Texas Instruments developed the first IC or integrated circuit in the 1950s and thereafter, Robert Noyce of Fairchild Semiconductor solved some practical problems of this integrated circuit.

## Inside the IC

When we think integrated circuits, little black chips are what come to mind. But what's inside that black box?

The guts of an integrated circuit, visible after removing the top.

The real "meat" to an IC is a complex layering of semiconductor wafers, copper, and other materials, which interconnect to form transistors, resistors or other components in a circuit. The cut and formed combination of these wafers is called a die.

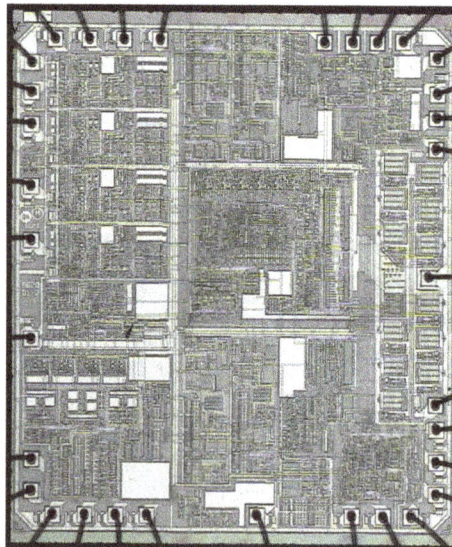

An overview of an IC die

While the IC itself is tiny, the wafers of semiconductor and layers of copper it consists of are incredibly thin. The connections between the layers are very intricate.

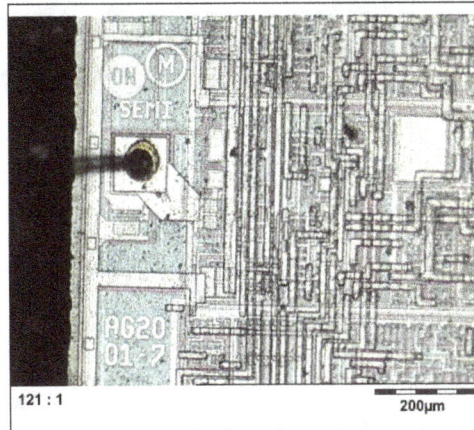

An IC die is the circuit in its smallest possible form, too small to solder or connect to. To make our job of connecting to the IC easier, we package the die. The IC package turns the delicate, tiny die, into the black chip we're all familiar with.

## IC Packages

The package is what encapsulates the integrated circuit die and splays it out into a device we can more easily connect to. Each outer connection on the die is connected via a tiny piece of gold wire to a pad or pin on the package. Pins are the silver, extruding terminals on an IC, which go on to connect to other parts of a circuit. These are of utmost importance to us, because they're what will go on to connect to the rest of the components and wires in a circuit.

There are many different types of packages, each of which has unique dimensions, mounting-types, and pin-counts.

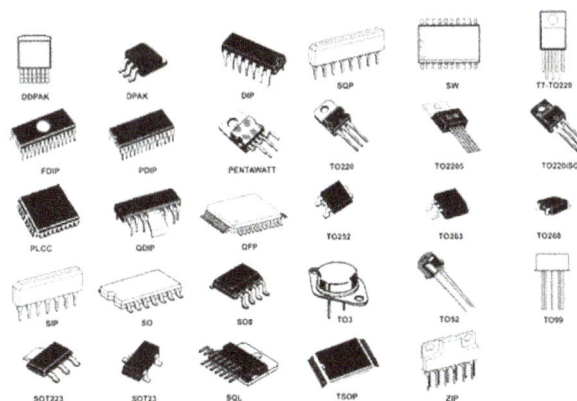

## Polarity Marking and Pin Numbering

All ICs are polarized, and every pin is unique in terms of both location and function. This means the package has to have some way to convey which pin is which. Most ICs will use either a notch or a dot to indicate which pin is the first pin. (Sometimes both, sometimes one or the other.)

Notch→

↑
Dot

Once you know where the first pin is, the remaining pin numbers increase sequentially as you move counter-clockwise around the chip.

16 15 14 13 12 11 10 9

Notch→

1 2 3 4 5 6 7 8

## Mounting Style

One of the main distinguishing package type characteristics is the way they mount to a circuit board. All packages fall into one of two mounting types: through-hole (PTH) or surface-mount (SMD or SMT). Through-hole packages are generally bigger, and much easier to work with. They're designed to be stuck through one side of a board and soldered to the other side.

Surface-mount packages range in size from small to minuscule. They are all designed to sit on one side of a circuit board and be soldered to the surface. The pins of a SMD package either extrude out the side, perpendicular to the chip, or are sometimes arranged in a matrix on the bottom of the chip. ICs in this form factor are not very "hand-assembly-friendly." They usually require special tools to aid in the process.

## DIP (Dual in-line packages)

DIP, short for dual in-line package, is the most common through-hole IC package you'll encounter. These little chips have two parallel rows of pins extending perpendicularly out of a rectangular, black, plastic housing.

The 28-pin ATmega328 is one of the more popular DIP-packaged microcontrollers.

Each of the pins on a DIP IC are spaced by 0.1" (2.54mm), which is a standard spacing and perfect for fitting into breadboards and other prototyping boards. The overall dimensions of a DIP package depend on its pin count, which may be anywhere from four to 64.

The area between each row of pins is perfectly spaced to allow DIP ICs to straddle the center area of a breadboard. This provides each of the pins its own row in the board, and it makes sure they don't short to each other.

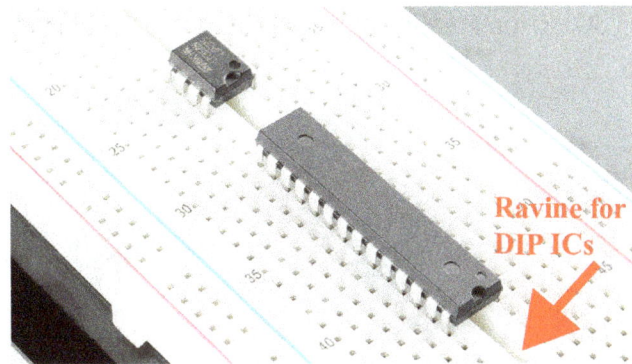

Aside from being used in breadboards, DIP ICs can also be soldered into PCBs. They're inserted into one side of the board and soldered into place on the other side. Sometimes, instead of soldering directly to the IC, it's a good idea to socket the chip. Using sockets allows for a DIP IC to be removed and swapped out, if it happens to "let its blue smoke out."

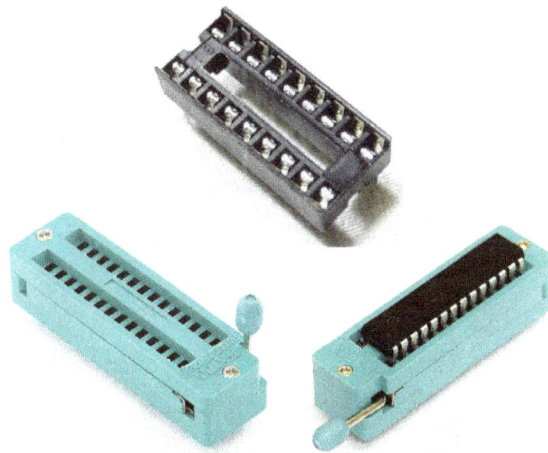

A regular DIP socket (top) and a ZIF socket with and without an IC.

## Surface-Mount (SMD/SMT) Packages

There is a huge variety of surface-mount package types these days. In order to work with surface-mount packaged ICs, you usually need a custom printed circuit board (PCB) made for them, which has a matching pattern of copper on which they're soldered.

Here are a few of the more common SMD package types out there, ranging in hand-solderability from "doable" to "doable, but only with special tools" to "doable only with very special, usually automated tools".

## Small-outline (SOP)

Small-outline IC (SOIC) packages are the surface-mount cousin of the DIP. It's what you'd get if you bent all the pins on a DIP outward, and shrunk it down to size. With a steady hand, and a close eye, these packages are among the easiest SMD parts to hand solder. On SOIC packages, each pin is usually spaced by about 0.05" (1.27mm) from the next.

The SSOP (shrink small-outline package) is an even smaller version of SOIC packages. Other, similar IC packages include TSOP (thin small-outline package) and TSSOP (thin-shrink small-outline package).

A 16-Channel Multiplexer (CD74HC4067) in a 24-pin SSOP package. Mounted on a board in the middle (quarter added for size-comparison).

A lot of the more simple, single-task-oriented ICs like the MAX232 or multiplexers come in SOIC or SSOP forms.

## Quad Flat Packages

Splaying IC pins out in all four directions gets you something that might look like a quad flat package (QFP). QFP ICs might have anywhere from eight pins per side (32 total) to upwards of seventy (300+ total). The pins on a QFP IC are usually spaced by anywhere from 0.4mm to 1mm. Smaller variants of the standard QFP package include thin (TQFP), very thin (VQFP), and low-profile (LQFP) packages.

The ATmega32U4 in a 44-pin (11 on each side) TQFP package

If you sanded the legs off a QFP IC, you get something that might look like a quad-flat no-leads (QFN) package. The connections on QFN packages are tiny, exposed pads on the bottom corner edges of the IC. Sometimes they wrap around, and are exposed on both the side and bottom, other packages only expose the pad on the bottom of the chip.

The multitalented MPU-6050 IMU sensor comes in a relatively tiny QFN package, with 24 total pins hiding on the bottom edge of the IC.

Thin (TQFN), very thin (VQFN), and micro-lead (MLF) packages are smaller variations of the standard QFN package. There are even dual no-lead (DFN) and thin-dual no-lead (TDFN) packages, which have pins on just two of the sides.

Many microprocessors, sensors, and other modern ICs come in QFP or QFN packages. The popular ATmega328microcontroller is offered in both a TQFP package and a QFN-type (MLF) form, while a tiny accelerometer/gyroscope like the MPU-6050 comes in a miniscule QFN form.

## Ball Grid Arrays

Finally, for really advanced ICs, there are ball grid array (BGA) packages. These are amazingly intricate little packages where little balls of solder are arranged in a 2-D grid on the bottom of the IC. Sometimes the solder balls are attached directly to the die!

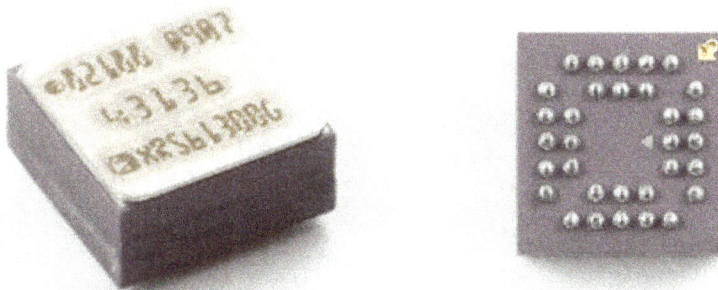

BGA packages are usually reserved for advanced microprocessors, like those on the pcDuino or Raspberry Pi.

If you can hand solder a BGA-packaged IC, consider yourself a master solderer. Usually, to put these packages onto a PCB requires an automated procedure involving pick-and-place machines and reflow ovens.

## Different Types of Integrated Circuits

There are different types of ICs; classification of Integrated Circuits is done based on various criteria. A few types of ICs in a system are shown in the below figure with their names in a tree format.

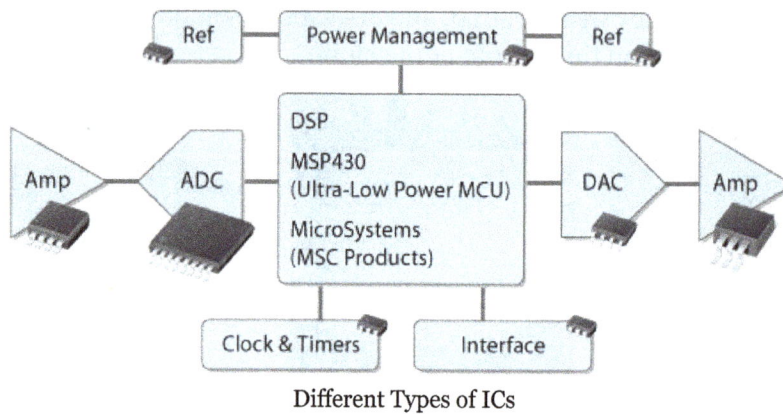

Different Types of ICs

Based on the intended application, the IC are classified as analog integrated circuits, digital integrated circuits and mixed integrated circuits.

## Digital Integrated Circuits

The integrated circuits that operate only at a few defined levels instead of operating over all levels of signal amplitude are called as Digital ICs and these are designed by using multiple numbers of digital logic gates, multiplexers, flip flops and other electronic components of circuits. These logic gates work with binary input data or digital input data, such as 0 (low or false or logic 0) and 1 (high or true or logic 1).

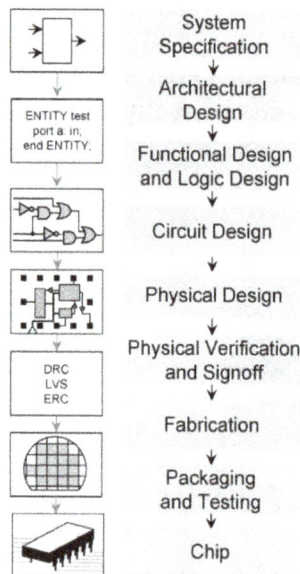

## Digital Integrated Circuits

The above figure shows the steps involved in designing a typical digital integrated circuits. These digital ICs are frequently used in the computers, microprocessors, digital signal processors, computer networks and frequency counters. There are different types of digital ICs or types of digital integrated circuits, such as programmable ICs, memory chips, logic ICs, power management ICs and interface ICs.

## Analog Integrated Circuits

The integrated circuits that operate over a continuous range of signal are called as Analog ICs. These are subdivided as linear Integrated Circuits (Linear ICs) and Radio Frequency Integrated Circuits (RF ICs). In fact, the relationship between the voltage and current maybe nonlinear in some cases over a long range of the continuous analog signal.

Analog Integrated Circuits

The frequently used analog IC is an operational amplifier or simply called as an op-amp, similar to the differential amplifier, but possesses a very high voltage gain. It consists of very less number of transistors compared to the digital ICs, and, for developing analog application specific integrated circuits (analog ASICs), computerized simulation tools are used.

## Mixed Integrated Circuits

The integrated circuits that are obtained by the combination of analog and digital ICs on a single chip are called as Mixed ICs. These ICs functions as Digital to Analog converters, Analog to Digital converters (D/A and A/D converters) and clock/timing ICs. The circuit depicted in the above figure is an example of mixed integrated circuit which is a photograph of the 8 to 18 GHz self-healing radar receiver.

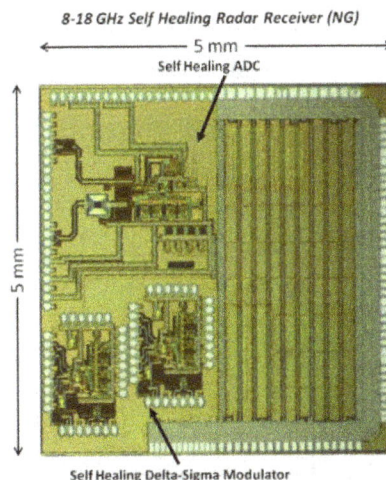

Mixed Integrated Circuits

This mixed-signal Systems-on-a-chip is a result of advances in the integration technology, which enabled to integrate digital, multiple analog and RF functions on a single chip.

General types of integrated circuits (ICs) include the following:

## Logic Circuits

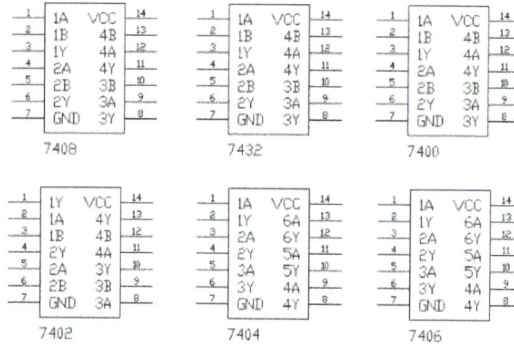

Logic Circuits

These ICs are designed using logic gates-that work with binary input and output (0 or 1). These are mostly used as decision makers. Based on the logic or truth table of the logic gates, all the logic gates connected in the IC give an output based on the circuit connected inside the IC- such that this output is used for performing a specific intended task. A few logic ICs are shown above.

## Comparators

Comparators

The comparator ICs are used as comparators for comparing the inputs and then to produce an output based on the ICs' comparison.

## Switching ICs

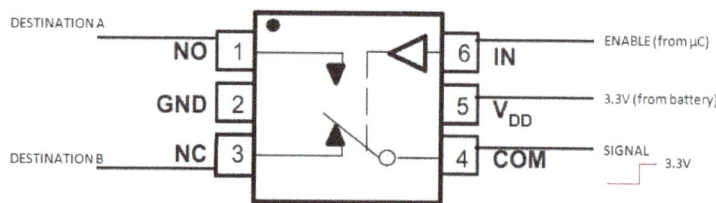

Switching ICs

Switches or Switching ICs are designed by using the transistors and are used for performing the switching operations. The above figure is an example showing an SPDT IC switch.

## Audio Amplifiers

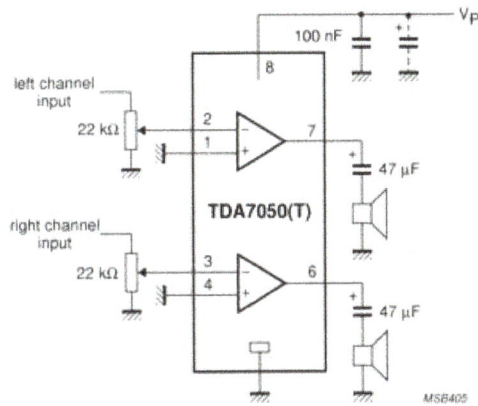

Audio amplifiers

The audio amplifiers are one of the many types of ICs, which are used for the amplification of the audio. These are generally used in the audio speakers, television circuits, and so on. The above circuit shows the low- voltage audio amplifier IC.

## Operational Amplifiers

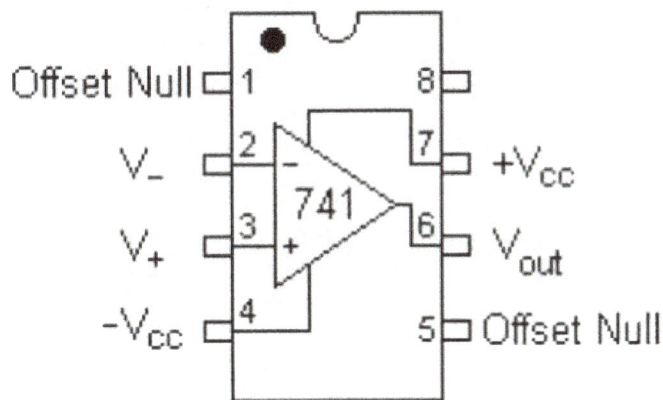

Operational amplifiers

The operational amplifiers are frequently used ICs, similar to the audio amplifiers which are used for the audio amplification. These op-amps are used for the amplification purpose, and these ICs work similar to the transistor amplifier circuits. The pin configuration of the 741 op-amp IC is shown in the above figure.

## Timer ICs

Timers are special purpose integrated circuits used for the purpose of counting and to keep a track of time in intended applications. The block diagram of the internal circuit of the LM555 timer IC is shown in the above circuit.

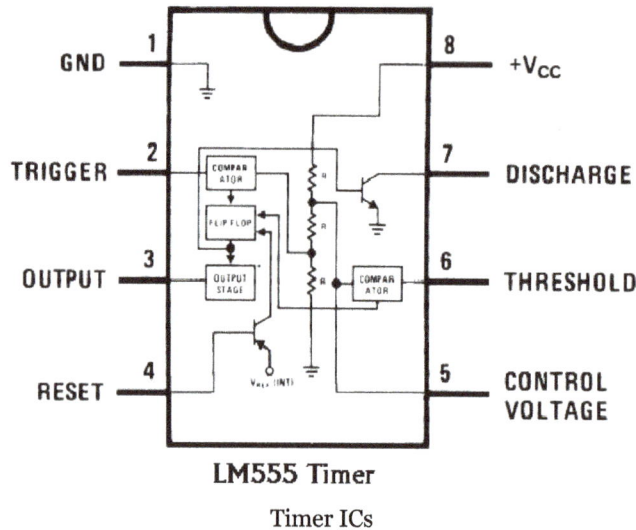

LM555 Timer

Timer ICs

Based on the number of components used (typically based on the number of transistors used), they are as follows:

- Small-scale integration consists of only a few transistors (tens of transistors on a chip), these ICs played a critical role in early aerospace projects.

- Medium-scale integration consists of some hundreds of transistors on the IC chip developed in the 1960s and achieved better economy and advantages compared to the SSI ICs.

- Large-scale integration consists of thousands of transistors on the chip with almost the same economy as medium scale integration ICs. The first microprocessor, calculator chips and RAMs of 1Kbit developed in the 1970s had below four thousand transistors.

- Very large-scale integration consists of transistors from hundreds to several billions in number. (Development period: from 1980s to 2009)

- Ultra large-scale integration consists of transistors in excess of more than one million, and later wafer-scale integration (WSI), system on a chip (SoC) and three dimensional integrated circuit (3D-IC) were developed.

All these can be treated as generations of integrated technology. ICs are also classified based on the fabrication process and packing technology. There are numerous types of ICs among which, an IC will function as timer, counter, register, amplifier, oscillator, logic gate, adder, microprocessor, and so on.

The conventional Integrated circuits are reduced in practical usage, because of the invention of the Nano-electronics and the miniaturization of ICs being continued by this Nano-electronics technology. However, the conventional ICs are not yet replaced by Nano-electronics but the usage of the conventional ICs is getting diminished partially.

## Applications of Integrated Circuits

The applications of an ICs includes the following:

- Radar

- Wristwatches

- Televisions

- Juice Makers

- PC

- Video Processors

- Audio Amplifiers

- Memory Devices

- Logic Devices

- Radio Frequency Encoders and Decoders

## Integrated Circuit Design

### Basic Semiconductor Design

Any material can be classified as one of three types: conductor, insulator, or semiconductor. A conductor (such as copper or salt water) can easily conduct electricity because it has an abundance of free electrons. An insulator (such as ceramic or dry air) conducts electricity very poorly because it has few or no free electrons. A semiconductor (such as silicon or gallium arsenide) is somewhere between a conductor and an insulator. It is capable of conducting some electricity, but not much.

### Doping silicon

Most ICs are made of silicon, which is abundant in ordinary beach sand. Pure crystalline silicon, as with other semiconducting materials, has a very high resistance to electrical current at normal room temperature. However, with the addition of certain impurities, known as dopants, the silicon can be made to conduct usable currents. In particular, the doped silicon can be used as a switch, turning current off and on as desired.

The process of introducing impurities is known as doping or implantation. Depending on a dopant's atomic structure, the result of implantation will be either an n-type (negative) or a p-type (positive) semiconductor. An n-type semiconductor results from implanting dopant atoms that have more electrons in their outer (bonding) shell than silicon. The resulting semiconductor crystal contains excess, or free, electrons that are available for conducting current. A p-type semiconductor results from implanting dopant atoms that have fewer electrons in their outer shell than silicon. The resulting crystal contains "holes" in its bonding structure where electrons would normally be located. In essence, such holes can move through the crystal conducting positive charges.

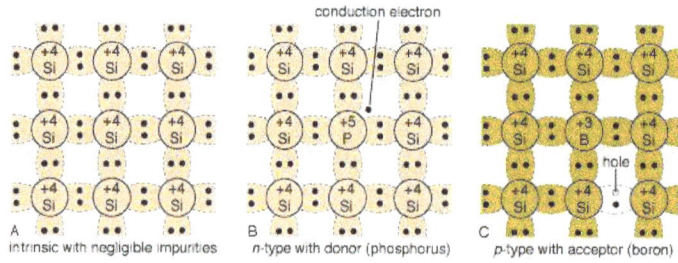

Three bond pictures of a semiconductor.

## The p-n Junction

A p-type or an n-type semiconductor is not very useful on its own. However, joining these opposite materials creates what is called a p-n junction. A p-n junction forms a barrier to conduction between the materials. Although the electrons in the n-type material are attracted to the holes in the p-type material, the electrons are not normally energetic enough to overcome the intervening barrier. However, if additional energy is provided to the electrons in the n-type material, they will be capable of crossing the barrier into the p-type material—and current will flow. This additional energy can be supplied by applying a positive voltage to the p-type material. The negatively charged electrons will then be highly attracted to the positive voltage across the junction.

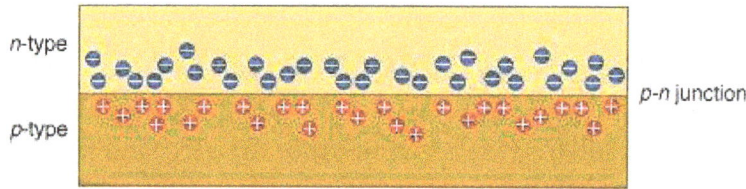

The p-n junction, a barrier forms along the boundary between p-type and n-type semiconductors that is known as a p-n junction. Because electrons under ordinary conditions will flow in only one direction through such barriers, p-n junctions form the basis for creating electronic rectifiers and switches.

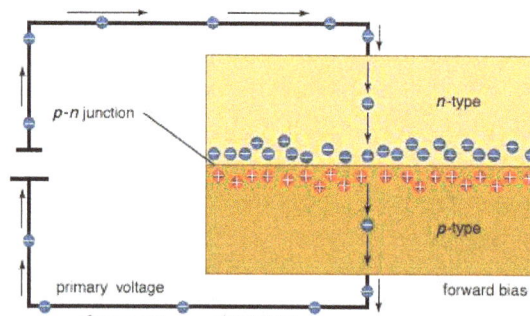

A forward-biased p-n junction adding a small primary voltage such that the electron source (negative terminal) is attached to the n-type semiconductor surface and the drain (positive terminal) is attached to the p-type semiconductor surface results in a small continuous current. This arrangement is referred to as being forward-biased

A p-n junction that conducts electricity when energy is added to the n material is called forward-biased because the electrons move forward into the holes. If voltage is applied in the opposite direction—a positive voltage connected to the n side of the junction—no current will flow. The electrons in the n material will still be attracted to the positive voltage, but the voltage will now be on the same side of the barrier as the electrons. In this state a junction is said to be reverse-biased.

Since p-n junctions conduct electricity in only one direction, they are a type of diode. Diodes are essential building blocks of semiconductor switches.

## Field-effect Transistors

Bringing a negative voltage close to the centre of a long strip of n-type material will repel nearby electrons in the material and thus form holes—that is, transform some of the strip in the middle to p-type material. This change in polarity using an electric field gives the field-effect transistor its name. While the voltage is being applied, there will exist two p-n junctions along the strip, from n to p and then from p back to n. One of the two junctions will always be reverse-biased. Since reverse-biased junctions cannot conduct, current cannot flow through the strip.

The field effect can be used to create a switch (transistor) to turn current off and on, simply by applying and removing a small voltage nearby in order to create or destroy reverse-biased diodes in the material. A transistor created by using the field effect is called a field-effect transistor (FET). The location where the voltage is applied is known as a gate. The gate is separated from the transistor strip by a thin layer of insulation to prevent it from short-circuiting the flow of electrons through the semiconductor from an input (source) electrode to an output (drain) electrode.

Similarly, a switch can be made by placing a positive gate voltage near a strip of p-type material. A positive voltage attracts electrons and thus forms a region of n within a strip of p. This again creates two p-n junctions, or diodes. As before, one of the diodes will always be reverse-biased and will stop current from flowing.

FETs are good for building logic circuits because they require only a small current during switching. No current is required for holding the transistor in an on or off state; a voltage will maintain the state. This type of switching helps preserve battery life. A FET is called unipolar (from "one polarity") because the main conduction method is either holes or electrons, not both.

## Enhancement-mode FETs

There are two basic types of FETs. The type described previously is a depletion-mode FET, since a region is depleted of its natural charge. The field effect can also be used to create what is called an enhancement-mode FET by enhancing a region to appear similar to its surrounding regions.

An n-type enhancement-mode FET is made from two regions of n-type material separated by a small region of p. As this FET naturally contains two p-n junctions—two diodes—it is normally switched off. However, when a positive voltage is placed on the gate, the voltage attracts electrons and creates n-type material in the middle region, filling the gap that was previously p-type material. The gate voltage thus creates a continuous region of n across the entire strip, allowing current to flow from one side to the other. This turns the transistor on. Similarly, a p-type enhancement-mode FET can be made from two regions of p-type material separated by a small region of n. The gate voltage required for turning on this transistor is negative. Enhancement-mode FETs switch faster than depletion-mode FETs because they require a change only near the surface under the gate, rather than all the way through the material.

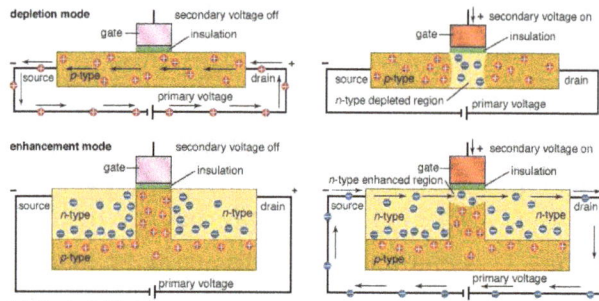

Depletion mode versus enhancement mode MOSFETsIn depletion mode metal-oxide-semiconductor field-effect transistors (MOSFETs), a secondary voltage is applied to deplete the region under the gate of charge carriers, thereby pinching off the current. In enhancement mode MOSFETs, a secondary voltage is used to enhance charge carriers beneath the gate, thereby allowing current to flow.

## Complementary Metal-oxide Semiconductors

Recall that placing a positive voltage at the gate of an n-type enhanced-mode FET will turn the switch on. Placing the same voltage at the gate of a p-type enhanced-mode FET will turn the switch off. Likewise, placing a negative voltage at the gate will turn the n-type off and the p-type on. These FETs always respond in opposite, or complementary, fashion to a given gate voltage. Thus, if the gates of an n-type and a p-type FET are connected any voltage applied to the common gate will operate the complementary pair, turning one on and leaving the other off. A semiconductor that pairs n- and p-type transistors this way is called a complementary metal-oxide semiconductor (CMOS). Because complementary transistor pairs can quickly switch between two logic states, CMOSs are very useful in logic circuits. In particular, because only one circuit is on at any time, CMOSs require less power and are often used for battery-powered devices, such as in digital cameras, and for the special memory that holds the date, time, and system parameters in personal computers.

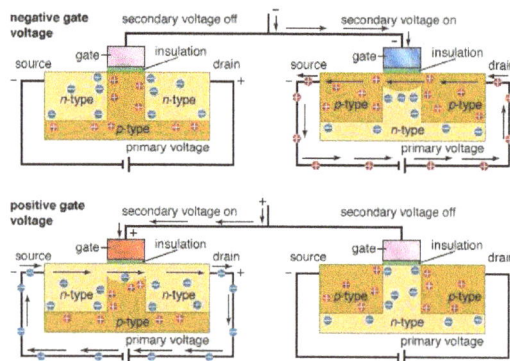

CMOSA complementary metal-oxide semiconductor (CMOS) consists of a pair of semiconductors connected to a common secondary voltage such that they operate in opposite (complementary) fashion. Thus, when one transistor is turned on, the other is turned off, and vice versa.

## Bipolar Transistors

Bipolar transistors simultaneously use holes and electrons to conduct, hence their name (from "two polarities"). Like FETs, bipolar transistors contain p- and n-type materials configured in input, middle, and output regions. In bipolar transistors, however, these regions are referred to as the emitter, the base, and the collector. Instead of relying, as FETs do, on a secondary voltage source to change the polarity beneath the gate (the field effect), bipolar transistors use a secondary voltage source to provide enough energy for electrons to punch through the reverse-biased base-collector junction. As the electrons are energized, they jump into the collector and complete the circuit. Note that even with highly energetic electrons, the middle section of p-type material must be extremely thin for the electrons to pass through both junctions.

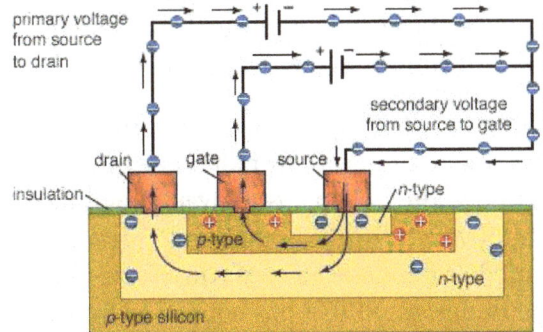

Bipolar transistor This type of transistor is called bipolar because both electrons and "holes" are used to carry charges through the n-p-n or p-n-p junction

A bipolar base region can be fabricated that is much smaller than any CMOS transistor gate. This smaller size enables bipolar transistors to operate much faster than CMOS transistors. Bipolar transistors are typically used in applications where speed is very important, such as in radio-frequency ICs. On the other hand, although bipolar transistors are faster, FETs use less current. The type of switch a designer selects depends on which benefits are more important for the application: speed or power savings. This is one of many trade-off decisions engineers make in designing their circuits.

## Designing ICs

All ICs use the same basic principles of voltage (V), current (I), and resistance (R). In particular, equations based on Ohm's law, V = IR, determine many circuit design choices. Design engineers must also be familiar with the properties of various electronic components needed for different applications.

## Analog Design

As mentioned earlier, an analog circuit takes an infinitely variable real-world voltage or current and modifies it in some useful way. The signal might be amplified, compared with another signal, mixed with other signals, separated from other signals, examined for value, or otherwise manipulated. For the design of this type of circuit, the choice of every individual component, size, placement, and connection is crucial. Unique decisions abound—for instance, whether one connection should be slightly wider than another connection, whether one resistor should be oriented parallel or perpendicular to another, or whether one wire can lie over the top of another. Every small detail affects the final performance of the end product.

When integrated circuits were much simpler, component values could be calculated by hand. For instance, a specific amplification value (gain) of an amplifier could typically be calculated from the ratio of two specific resistors. The current in the circuit could then be determined, using the resistor value required for the amplifier gain and the supply voltage used. As designs became more complex, laboratory measurements were used to characterize the devices. Engineers drew graphs of device characteristics across several variables and then referred to those graphs as they needed information for their calculations. As scientists improved their characterization of the intricate physics of each device, they developed complex equations that took into account subtle effects that were not apparent from coarse laboratory measurements. For example, a transistor works very differently at different frequencies, sizes, orientations, and placements. In particular, scientists found parasitic components (unwanted effects, usually resistance and capacitance) that are inherent in the way the devices are built. Parasitics become more problematic as the circuitry becomes more sophisticated and smaller and as it runs at higher frequencies.

Although parasitic components in a circuit can now be accounted for by sophisticated equations, such calculations are very time-consuming to do by hand. For this work computers have become indispensable. In particular, a public-domain circuit-analysis program developed at the University of California, Berkeley, during the 1970s, SPICE (Simulation Program with Integrated Circuit Emphasis), and various proprietary models designed for use with it are ubiquitous in engineering courses and in industry for analog circuit design. SPICE has equations for transistors, capacitors, resistors, and other components, as well as for lengths of wires and for turns in wires, and it can reduce the calculation of circuit interactions to hours from the months formerly required for hand calculations.

## Digital Design

Since digital circuits involve millions of times as many components as analog circuits, much of the design work is done by copying and reusing the same circuit functions, especially by using digital design software that contains libraries of pre structured circuit components. The components available in such a library are of similar height, contain contact points in predefined locations, and have other rigid conformities so that they fit together regardless of how the computer configures a layout. While SPICE is perfectly adequate for analyzing analog circuits, with equations that describe individual components, the complexity of digital circuits requires a less-detailed approach. Therefore, digital analysis software ignores individual components for mathematical models of entire preconfigured circuit blocks (or logic functions).

Whether analog or digital circuitry is used depends on the function of a circuit. The design and layout of analog circuits are more demanding of teamwork, time, innovation, and experience, particularly as circuit frequencies get higher, though skilled digital designers and layout engineers can be of great benefit in overseeing an automated process as well. Digital design emphasizes different skills from analog design.

## Mixed-signal Design

For designs that contain both analog and digital circuitry (mixed-signal chips), standard analog and digital simulators are not sufficient. Instead, special behavioral simulators are used, employing the same simplifying idea behind digital simulators to model entire circuits rather than individual transistors. Behavioral simulators are designed primarily to speed up simulations of the analog side of a mixed-signal chip.

The difficulty with behavioral simulation is making sure that the model of the analog circuit function is accurate. Since each analog circuit is unique, it seems as though one must design the system twice—once to design the circuitry and once to design the model for the simulator.

## Raw Materials

Pure silicon is the basis for most integrated circuits. It provides the base, or substrate for the entire chip and is chemically doped to provide the N and P regions that make up the integrated circuit components. The silicon must be so pure that only one out of every ten billion atoms can be an impurity. This would be the equivalent of one grain of sugar in ten buckets of sand. Silicon dioxide is used as an insulator and as a dielectric material in IC capacitors.

Typical N-type dopants include phosphorus and arsenic. Boron and gallium are typical P-type dopants. Aluminum is commonly used as a connector between the various IC components. The thin wire leads from the integrated circuit chip to its mounting package may be aluminum or gold. The mounting package itself may be made from ceramic or plastic materials.

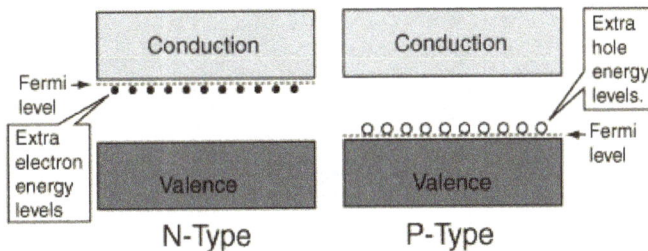

## Manufacturing Process

Hundreds of integrated circuits are made at the same time on a single, thin slice of silicon and are then cut apart into individual IC chips. The manufacturing process takes place in a tightly controlled environment known as a clean room where the air is filtered to remove foreign particles. The few equipment operators in the room wear lint-free garments, gloves, and coverings for their heads and feet. Since some IC components are sensitive to certain frequencies of light, even the light sources are filtered. Although manufacturing processes may vary depending on the integrated circuit being made, the following process is typical.

### Preparing the Silicon Wafer

- A cylindrical ingot of silicon about 1.5 to 4.0 inches (3.8 to 10.2 cm) in diameter is held vertically inside a vacuum chamber with a high-temperature heating coil encircling it. Starting at the top of the cylinder, the silicon is heated to its melting point of about 2550° F (1400° C). To avoid contamination, the heated region is contained only by the surface tension of the molten silicon. As the region melts, any impurities in the silicon become mobile. The heating coil is slowly moved down the length of the cylinder, and the impurities are carried along with the melted region. When the heating coil reaches the bottom, almost all of the impurities have been swept along and are concentrated there. The bottom is then sliced off, leaving a cylindrical ingot of purified silicon.

- A thin, round wafer of silicon is cut off the ingot using a precise cutting machine called a wafer slicer. Each slice is about 0.01 to 0.025 inches (0.004 to 0.01 cm) thick. The surface on which the integrated circuits are to be formed is polished.

- The surfaces of the wafer are coated with a layer of silicon dioxide to form an insulating base and to prevent any oxidation of the silicon which would cause impurities. The silicon dioxide is formed by subjecting the wafer to superheated steam at about 1830° F (1000° C) under several atmospheres of pressure to allow the oxygen in the water vapor to react with the silicon. Controlling the temperature and length of exposure controls the thickness of the silicon dioxide layer.

## Basic Process Steps for Wafer Preparation

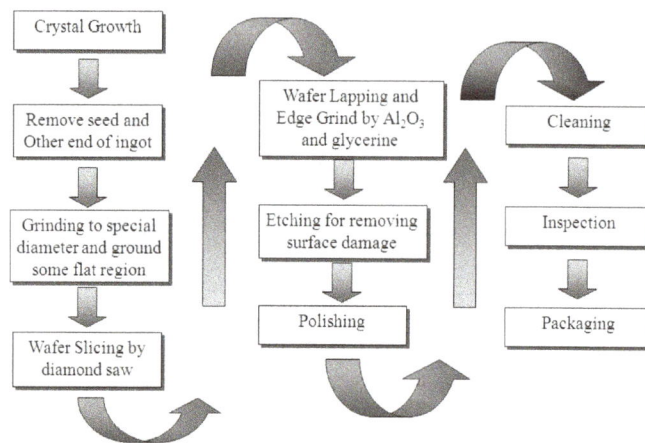

## Masking

- The complex and interconnected design of the circuits and components is prepared in a process similar to that used to make printed circuit boards. For ICs, however, the dimensions are much smaller and there are many layers superimposed on top of each other. The design of each layer is prepared on a computer-aided drafting machine, and the image is made into a mask which will be optically reduced and transferred to the surface of the wafer. The mask is opaque in certain areas and clear in others. It has the images for all of the several hundred integrated circuits to be formed on the wafer.

- A drop of photoresist material is placed in the center of the silicon wafer, and the wafer is spun rapidly to distribute the photoresist over the entire surface. The photoresist is then baked to remove the solvent.

- The coated wafer is then placed under the first layer mask and irradiated with light. Because the spaces between circuits and components are so small, ultraviolet light with a very short wavelength is used to squeeze through the tiny clear areas on the mask. Beams of electrons or x-rays are also sometimes used to irradiate the photoresist.

- The mask is removed and portions of the photoresist are dissolved. If a positive photoresist was used, then the areas that were irradiated will be dissolved. If a negative photoresist was used, then the areas that were irradiated will remain. The uncovered areas are then either chemically etched to open up a layer or is subjected to chemical doping to create a layer of P or N regions.

## Doping Atomic Diffusion

- One method of adding dopants to create a layer of P or N regions is atomic diffusion. In this method a batch of wafers is placed in an oven made of a quartz tube surrounded by a heating element. The wafers are heated to an operating temperature of about 1500-2200° F (816-1205° C), and the dopant chemical is carried in on an inert gas. As the dopant and gas pass over the wafers, the dopant is deposited on the hot surfaces left exposed by the masking process.

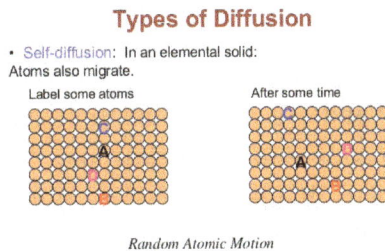

**Types of Diffusion**

- Self-diffusion: In an elemental solid:
Atoms also migrate.

Label some atoms          After some time

*Random Atomic Motion*

This method is good for doping relatively large areas, but is not accurate for smaller areas. There are also some problems with the repeated use of high temperatures as successive layers are added.

## Doping Ion Implantation

- The second method to add dopants is ion implantation. In this method a dopant gas, like phosphine or boron trichloride, is ionized to provide a beam of high-energy dopant ions which are fired at specific regions of the wafer. The ions penetrate the wafer and remain implanted. The depth of penetration can be controlled by altering the beam energy, and the amount of dopant can be controlled by altering the beam current and time of exposure. Schematically, the whole process resembles firing a beam in a bent cathode-ray tube. This method is so precise, it does not require masking—it just points and shoots the dopant where it is needed. However it is much slower than the atomic diffusion process.

## Making Successive Layers

- The process of masking and etching or doping is repeated for each successive layer depending on the doping process used until all of the integrated circuit chips are complete. Sometimes a layer of silicon dioxide is laid down to provide an insulator between layers or components. This is done through a process known as chemical vapor deposition, in which the wafer's surface is heated to about 752° F (400° C), and a reaction between the gases silane and oxygen deposits a layer of silicon dioxide. A final silicon dioxide layer seals the surface, a final etching opens up contact points, and a layer of aluminum is deposited to make the contact pads. At this point, the individual ICs are tested for electrical function.

### Making Individual ICs

- The thin wafer is like a piece of glass. The hundreds of individual chips are separated by scoring a crosshatch of lines with a fine diamond cutter and then putting the wafer under stress to cause each chip to separate. Those ICs that failed the electrical test are discarded. Inspection under a microscope reveals other ICs that were damaged by the separation process, and these are also discarded.

- The good ICs are individually bonded into their mounting package and the thin wire leads are connected by either ultrasonic bonding or thermo compression. The mounting package is marked with identifying part numbers and other information.

- The completed integrated circuits are sealed in anti-static plastic bags to be stored or shipped to the end user.

### Quality Control

Despite the controlled environment and use of precision tools, a high number of integrated circuit chips are rejected. Although the percentage of reject chips has steadily dropped over the years, the task of making an interwoven lattice of microscopic circuits and components is still difficult, and a certain amount of rejects are inevitable.

### Hazardous Materials and Recycling

The dopants gallium and arsenic, among others, are toxic substances and their storage, use, and disposal must be tightly controlled.

Because integrated circuit chips are so versatile, a significant recycling industry has sprung up. Many ICs and other electronic components are removed from otherwise obsolete equipment, tested, and resold for use in other devices.

## Very Large Scale Integration

Very large-scale integration (VLSI) is the process of integrating or embedding hundreds of thousands of transistors on a single silicon semiconductor microchip. VLSI technology was conceived in the late 1970s when advanced level computer processor microchips were under development.

VLSI is a successor to large-scale integration (LSI), medium-scale integration (MSI) and small-scale integration (SSI) technologies.

VLSI is one of the most widely used technologies for microchip processors, integrated circuits (IC) and component designing. It was initially designed to support hundreds of thousands of transistor gates on a microchip which, as of 2012, exceeded several billion. All of these transistors are remarkably integrated and embedded within a microchip that has shrunk over time but still has the capacity to hold enormous amounts of transistors.

The first 1 megabyte RAM was built on top of VLSI design principles and included more than one million transistors on its microchip dye.

## References

- Integrated-circuits: sparkfun.com, Retrieved 11 April 2018

- Different-types-of-integrated-circuits: elprocus.com, Retrieved 15 May 2018

- How-integrated-circuits-work-physically: elprocus.com, Retrieved 25 March 2018

- Integrated-circuit-236561: britannica.com, Retrieved 30 May 2018

- Very-large-scale-integration-vlsi-714: techopedia.com, Retrieved 19 June 2018

# 2

# Logic Gates and Logic Synthesis

In electronics, a logic gate is a device implementing a Boolean function or performing a logical function on a set of binary inputs and performing a single output. These are constructed using diodes or transistors. This chapter closely examines the fundamentals of logic gates and logic synthesis. It includes vital topics related to combinatorial logic, CMOS, logic synthesis, VLSI design, interconnect delay, combinational network delay, etc.

## Logic Gates

Logic gates process signals which represent true or false. Normally the positive supply voltage +Vs represent true and oV represents false. Other terms used for the true and false states are shown in the table, it is best to be familiar with them all.

Gates are identified by their function: NOT, AND, NAND, OR, NOR, EX-OR and EX-NOR. Capital letters are normally used to make it clear that the term refers to a logic gate.

| Logic states | |
|---|---|
| True | False |
| 1 | 0 |
| High | Low |
| +Vs | oV |
| On | Off |

Note that logic gates are not always required because simple logic functions can be performed with switches or diodes, for example:

- Switches in series (AND function)
- Switches in parallel (OR function)
- Combining IC outputs with diodes (OR function)

A Digital Logic Gate is an electronic device that makes logical decisions based on the different combinations of digital signals present on its inputs.

Digital logic gates may have more than one input, (A, B, C, etc.) but generally only have one digital output, (Q). Individual logic gates can be connected together to form combinational or sequential circuits, or larger logic gate functions.

Standard commercially available digital logic gates are available in two basic families or forms, TTL which stands for Transistor-Transistor Logic such as the 7400 series, and CMOSwhich stands for Complementary Metal-Oxide-Silicon which is the 4000 series of chips. This notation of TTL or CMOS refers to the logic technology used to manufacture the integrated circuit, (IC) or a "chip" as it is more commonly called.

## Digital Logic Gate

Generally speaking, TTL logic IC's use NPN and PNP type Bipolar Junction Transistors while CMOS logic IC's use complementary MOSFET or JFET type Field Effect Transistors for both their input and output circuitry.

As well as TTL and CMOS technology, simple digital logic gates can also be made by connecting together diodes, transistors and resistors to produce RTL, Resistor-Transistor logic gates, DTL, Diode-Transistor logic gates or ECL, Emitter-Coupled logic gates but these are less common now compared to the popular CMOS family.

Integrated Circuits or IC's as they are more commonly called, can be grouped together into families according to the number of transistors or "gates" that they contain. For example, a simple AND gate my contain only a few individual transistors, were as a more complex microprocessor may contain many thousands of individual transistor gates. Integrated circuits are categorised according to the number of logic gates or the complexity of the circuits within a single chip with the general classification for the number of individual gates given as:

## Classification of Integrated Circuits

- Small Scale Integration or (SSI) – Contain up to 10 transistors or a few gates within a single package such as AND, OR, NOT gates.

- Medium Scale Integration or (MSI) – Between 10 and 100 transistors or tens of gates within a single package and perform digital operations such as adders, decoders, counters, flip-flops and multiplexers.

- Large Scale Integration or (LSI) – Between 100 and 1,000 transistors or hundreds of gates and perform specific digital operations such as I/O chips, memory, arithmetic and logic units.

- Very-Large Scale Integration or (VLSI) – Between 1,000 and 10,000 transistors or thousands of gates and perform computational operations such as processors, large memory arrays and programmable logic devices.

- Super-Large Scale Integration or (SLSI) – Between 10,000 and 100,000 transistors within a single package and perform computational operations such as microprocessor chips, micro-controllers, basic PICs and calculators.

- Ultra-Large Scale Integration or (ULSI) – More than 1 million transistors – the big boys that are used in computers CPUs, GPUs, video processors, micro-controllers, FPGAs and complex PICs.

While the "ultra large scale" ULSI classification is less well used, another level of integration which represents the complexity of the Integrated Circuit is known as the System-on-Chip or (SOC) for short. Here the individual components such as the microprocessor, memory, peripherals, I/O logic etc, are all produced on a single piece of silicon and which represents a whole electronic system within one single chip, literally putting the word "integrated" into integrated circuit.

These complete integrated chips which can contain up to 100 million individual silicon-CMOS transistor gates within one single package are generally used in mobile phones, digital cameras, micro-controllers, PIC's and robotic type applications.

## Moore's Law

In 1965, Gordon Moore co-founder of the Intel corporation predicted that "The number of transistors and resistors on a single chip will double every 18 months" regarding the development of semiconductor gate technology. When Gordon Moore made his famous comment way back in 1965 there were approximately only 60 individual transistor gates on a single silicon chip or die.

The worlds first microprocessor in 1971 was the Intel 4004 that had a 4-bit data bus and contained about 2,300 transistors on a single chip, operating at about 600kHz. Today, the Intel Corporation have placed a staggering 1.2 Billion individual transistor gates onto its new Quad-core i7-2700K Sandy Bridge 64-bit microprocessor chip operating at nearly 4GHz, and the on-chip transistor count is still rising, as newer faster microprocessors and micro-controllers are developed.

## Digital Logic States

The Digital Logic Gate is the basic building block from which all digital electronic circuits and microprocessor based systems are constructed from. Basic digital logic gates perform logical operations of AND, OR and NOT on binary numbers.

In digital logic design only two voltage levels or states are allowed and these states are generally referred to as Logic "1" and Logic "0", High and Low, or True and False. These two states are represented in Boolean Algebra and standard truth tables by the binary digits of "1" and "0" respectively.

A good example of a digital state is a simple light switch as it is either "ON" or "OFF" but not both at the same time. Then we can summarise the relationship between these various digital states as being:

| Boolean Algebra | Boolean Logic | Voltage State |
|:---:|:---:|:---:|
| Logic "1" | True (T) | High (H) |
| Logic "0" | False (F) | Low (L) |

Most digital logic gates and digital logic systems use "Positive logic", in which a logic level "0" or "LOW" is represented by a zero voltage, 0v or ground and a logic level "1" or "HIGH" is represented by a higher voltage such as +5 volts, with the switching from one voltage level to the other, from either a logic level "0" to a "1" or a "1" to a "0" being made as quickly as possible to prevent any faulty operation of the logic circuit.

There also exists a complementary "Negative Logic" system in which the values and the rules of a logic "0" and a logic "1" are reversed. Digital logic gates we shall only refer to the positive logic convention as it is the most commonly used.

In standard TTL (transistor-transistor logic) IC's there is a pre-defined voltage range for the input and output voltage levels which define exactly what is a logic "1" level and what is a logic "0" level and these are shown below.

## TTL Input and Output Voltage Levels

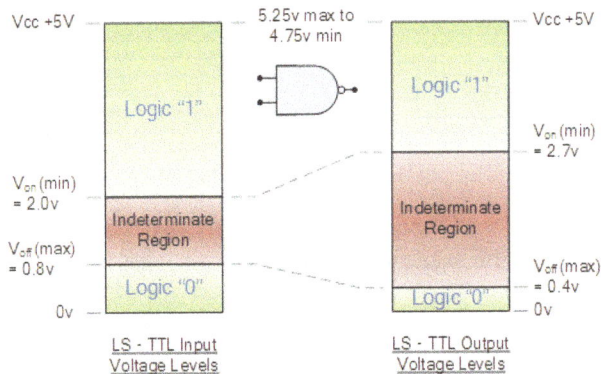

There are a large variety of logic gate types in both the bipolar 7400 and the CMOS 4000 families of digital logic gates such as 74Lxx, 74LSxx, 74ALSxx, 74HCxx, 74HCTxx, 74ACTxx etc, with each one having its own distinct advantages and disadvantages compared to the other. The exact switching voltage required to produce either a logic "0" or a logic "1" depends upon the specific logic group or family.

However, when using a standard +5 volt supply any TTL voltage input between 2.0v and 5v is considered to be a logic "1" or "HIGH" while any voltage input below 0.8v is recognised as a logic "0" or "LOW". The voltage region in between these two voltage levels either as an input or as an output is called the Indeterminate Region and operating within this region may cause the logic gate to produce a false output.

The CMOS 4000 logic family uses different levels of voltages compared to the TTL types as they are designed using field effect transistors, or FET's. In CMOS technology a logic "1" level operates between 3.0 and 18 volts and a logic "0" level is below 1.5 volts. Then the following table shows the difference between the logic levels of traditional TTL and CMOS logic gates.

## TTL and CMOS Logic Levels

| Device Type | Logic 0 | Logic 1 |
|-------------|---------|---------|
| TTL | 0 to 0.8v | 2.0 to 5v ($V_{cc}$) |
| CMOS | 0 to 1.5v | 3.0 to 18v ($V_{DD}$) |

Then from the above observations, we can define the ideal TTL digital logic gate as one that has a "LOW" level logic "0" of 0 volts (ground) and a "HIGH" level logic "1" of +5 volts and this can be demonstrated as:

## Ideal TTL Digital Logic Gate Voltage Levels

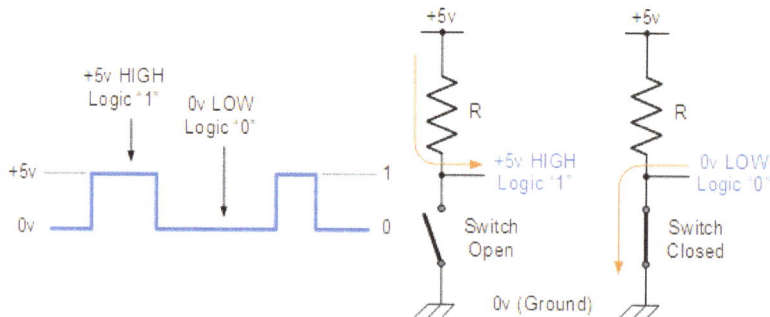

Where the opening or closing of the switch produces either a logic level "1" or a logic level "0" with the resistor R being known as a "pull-up" resistor.

## Digital Logic Noise

However, between these defined HIGH and LOW values lies what is generally called a "no-man's land" (the blue area's above) and if we apply a signal voltage of a value within this no-man's land area we do not know whether the logic gate will respond to it as a level "0" or as a level "1", and the output will become unpredictable.

Noise is the name given to a random and unwanted voltage that is induced into electronic circuits by external interference, such as from nearby switches, power supply fluctuations or from wires and other conductors that pick-up stray electromagnetic radiation. Then in order for a logic gate not to be influence by noise in must have a certain amount of noise margin or noise immunity.

## Digital Logic Gate Noise Immunity

The noise signal is superimposed onto the Vcc supply voltage and as long as it stays above the minimum level (VON(min)) the input an corresponding output of the logic gate are unaffected. But when the noise level becomes large enough and a noise spike causes the HIGH voltage level to drop below this minimum level, the logic gate may interpret this spike as a LOW level input and switch the output accordingly producing a false output switching. Then in order for the logic gate not to be affected by noise it must be able to tolerate a certain amount of unwanted noise on its input without changing the state of its output.

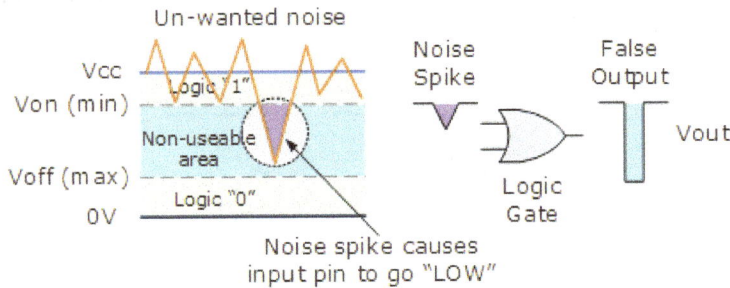

## Simple Basic Digital Logic Gates

Simple digital logic gates can be made by combining transistors, diodes and resistors with a simple example of a Diode-Resistor Logic (DRL) AND gate and a Diode-Transistor Logic (DTL) NAND gate given below:

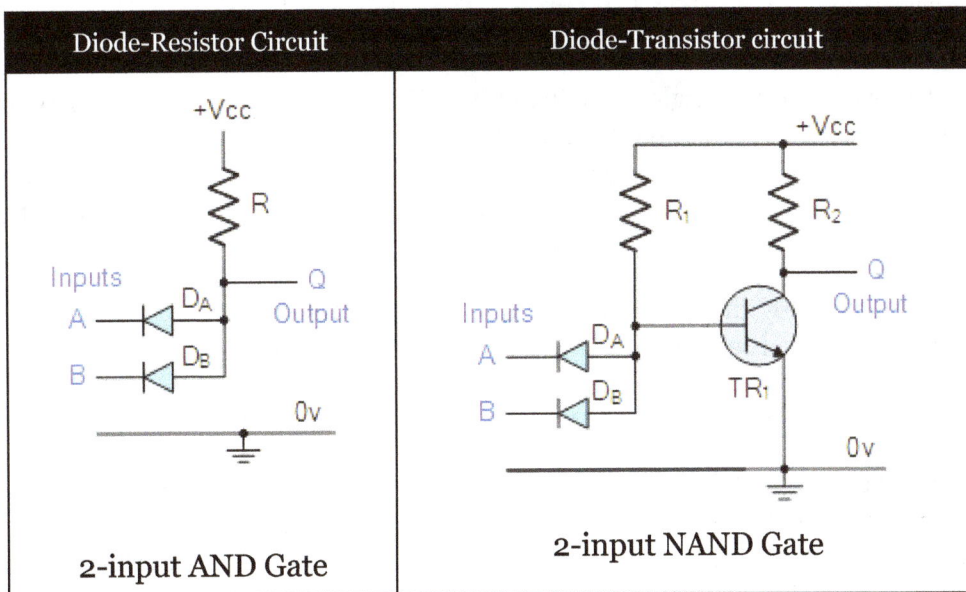

The simple 2-input Diode-Resistor AND gate can be converted into a NAND gate by the addition of a single transistor inverting (NOT) stage. Using discrete components such as diodes, resistors and transistors to make digital logic gate circuits are not used in practical commercially available logic IC's as these circuits suffer from propagation delay or gate delay and also power loss due to the pull-up resistors.

The disadvantage of diode-resistor logic is that there is no "Fan-out" facility which is the ability of a single output to drive many inputs of the next stages. Also this type of design does not turn fully "OFF" as a Logic "0" produces an output voltage of 0.6v (diode voltage drop), so the following TTL and CMOS circuit designs are used instead.

## Basic TTL Logic Gates

The simple Diode-Resistor AND gate above uses separate diodes for its inputs, one for each input. As a transistor is made up off two diode circuits connected together representing an NPN or a PNP device, the input diodes of the DTL circuit can be replaced by one single NPN transistor with multiple emitter inputs as shown.

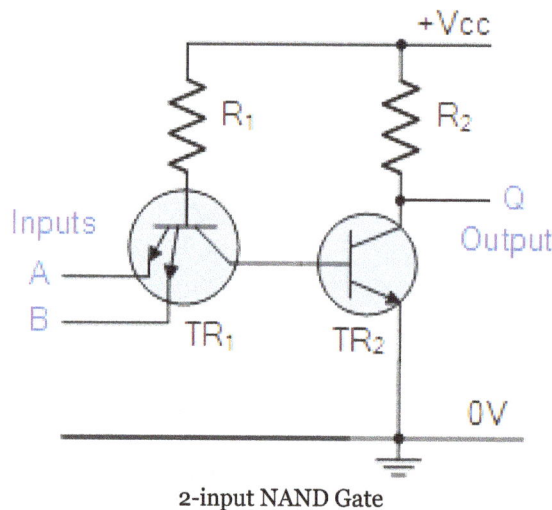

2-input NAND Gate

As the NAND gate contains a single stage inverting NPN transistor circuit (TR2) an output logic level "1" at Q is only present when both the emitters of TR1 are connected to logic level "0" or ground allowing base current to pass through the PN junctions of the emitter and not the collector. The multiple emitters of TR1 are connected as inputs thus producing a NAND gate function.

In standard TTL logic gates, the transistors operate either completely in the "cut off" region, or else completely in the saturated region, Transistor as a Switch type operation.

## Emitter Coupled Digital Logic Gate

Emitter Coupled Logic or ECL is another type of digital logic gate that uses bipolar transistor logic where the transistors are not operated in the saturation region, as they are with the standard TTL digital logic gate. Instead the input and output circuits are push-pull connected transistors with the supply voltage negative with respect to ground.

This has the effect of increasing the speed of operation of the emitter coupled logic gates up to the Gigahertz range compared with the standard TTL types, but noise has a greater effect in ECL logic, because the unsaturated transistors operate within their active region and amplify as well as switch signals.

## The "74" Sub-families of Integrated Circuits

With improvements in the circuit design to take account of propagation delays, current consumption, fan-in and fan-out requirements etc, this type of TTL bipolar transistor technology forms the basis of the prefixed "74" family of digital logic IC's, such as the "7400" Quad 2-input AND gate, or the "7402" Quad 2-input OR gate, etc.

Sub-families of the 74xx series IC's are available relating to the different technologies used to fabricate the gates and they are denoted by the letters in between the 74 designation and the device number. There are a number of TTL sub-families available that provide a wide range of switching speeds and power consumption such as the 74L00 or 74ALS00 AND gate, were the "L" stands for "Low-power TTL" and the "ALS" stands for "Advanced Low-power Schottky TTL" and these are listed below:

- 74xx or 74Nxx: Standard TTL – These devices are the original TTL family of logic gates introduced in the early 70's. They have a propagation delay of about 10ns and a power consumption of about 10mW. Supply voltage range: 4.75 to 5.25 volts.

- 74Lxx: Low Power TTL – Power consumption was improved over standard types by increasing the number of internal resistances but at the cost of a reduction in switching speed. Supply voltage range: 4.75 to 5.25 volts.

- 74Hxx: High Speed TTL – Switching speed was improved by reducing the number of internal resistances. This also increased the power consumption. Supply voltage range: 4.75 to 5.25 volts.

- 74Sxx: Schottky TTL – Schottky technology is used to improve input impedance, switching speed and power consumption (2mW) compared to the 74Lxx and 74Hxx types. Supply voltage range: 4.75 to 5.25 volts.

- 74LSxx: Low Power Schottky TTL – Same as 74Sxx types but with increased internal resistances to improve power consumption. Supply voltage range: 4.75 to 5.25 volts.

- 74ASxx: Advanced Schottky TTL – Improved design over 74Sxx Schottky types optimised to increase switching speed at the expense of power consumption of about 22mW. Supply voltage range: 4.5 to 5.5 volts.

- 74ALSxx: Advanced Low Power Schottky TTL – Lower power consumption of about 1mW and higher switching speed of 4nS compared to 74LSxx types. Supply voltage range: 4.5 to 5.5 volts.

- 74HCxx: High Speed CMOS – CMOS technology and transistors to reduce power consumption of less than 1uA with CMOS compatible inputs. Supply voltage range: 4.5 to 5.5 volts.

- 74HCTxx: High Speed CMOS – CMOS technology and transistors to reduce power consumption of less than 1uA but has increased propagation delay of about 16nS due to the TTL compatible inputs. Supply voltage range: 4.5 to 5.5 volts.

## Logic Gate Symbols

There are two series of symbols for logic gates. The traditional symbols have distinctive shapes making them easy to recognise, they are widely used in industry and education. The IEC (International Electrotechnical Commission) symbols are rectangles with a symbol inside to show the gate function. They are rarely used despite their official status but you may need to know them for an examination.

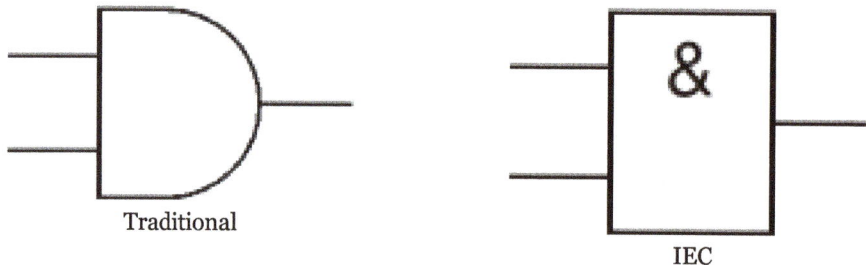

Traditional

IEC

## Inputs and Outputs

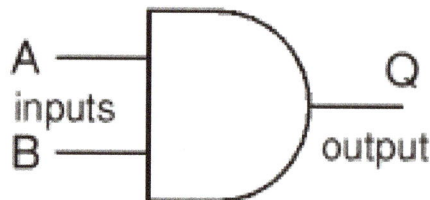

A inputs B — output Q

Gates have two or more inputs, except a NOT gate which has only one input. All gates have only one output. Usually the letters A, B, C and so on are used to label inputs, and Q is used to label the output. On this page the inputs are shown on the left and the output on the right.

## Inverting Circle (O)

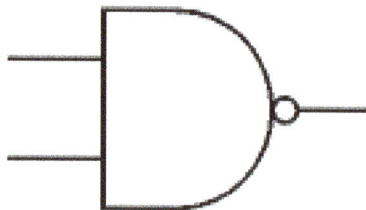

Some gate symbols have a circle on their output which means that their function includes inverting of the output. It is equivalent to feeding the output through a NOT gate. For example the NAND (Not-AND) gate symbol shown on the right is the same as an AND gate symbol but with the addition of an inverting circle on the output.

## Truth Tables

A truth table is a good way to show the function of a logic gate. It shows the output states for every possible combination of input states. The symbols 0 (false) and 1 (true) are usually used in truth tables. The example truth table shows the inputs and output of an AND gate.

There are summary truth tables below showing the output states for all types of 2-input and 3-input gates. These can be helpful if you are trying to select a suitable gate.

| Input A | Input B | Output Q |
|---------|---------|----------|
| 0 | 0 | 0 |
| 0 | 1 | 0 |
| 1 | 0 | 0 |
| 1 | 1 | 1 |

Summary truth tables

These summary truth tables below show the output states for all types of 2-input and 3-input gates. Note that EX-OR and EX-NOR gates can only have 2 inputs.

| Summary for all 2-input gates | | | | | | | |
|---|---|---|---|---|---|---|---|
| Inputs | | Output of each gate | | | | | |
| A | B | AND | NAND | OR | NOR | EX-OR | EX-NOR |
| 0 | 0 | 0 | 1 | 0 | 1 | 0 | 1 |
| 0 | 1 | 0 | 1 | 1 | 0 | 1 | 0 |
| 1 | 0 | 0 | 1 | 1 | 0 | 1 | 0 |
| 1 | 1 | 1 | 0 | 1 | 0 | 0 | 1 |

| Summary for all 3-input gates | | | | | | |
|---|---|---|---|---|---|---|
| Inputs | | | Output of each gate | | | |
| A | B | C | AND | NAND | OR | NOR |
| 0 | 0 | 0 | 0 | 1 | 0 | 1 |
| 0 | 0 | 1 | 0 | 1 | 1 | 0 |
| 0 | 1 | 0 | 0 | 1 | 1 | 0 |
| 0 | 1 | 1 | 0 | 1 | 1 | 0 |
| 1 | 0 | 0 | 0 | 1 | 1 | 0 |
| 1 | 0 | 1 | 0 | 1 | 1 | 0 |
| 1 | 1 | 0 | 0 | 1 | 1 | 0 |
| 1 | 1 | 1 | 1 | 0 | 1 | 0 |

## Logic ICs

Logic gates are available on ICs (chips) which usually contain several gates of the same type, for example the 4001 IC contains four 2-input NOR gates. There are several families of logic ICs and they can be split into two groups: the 4000 series and the 74 Series

The 4000 and 74HC families are the best for battery powered projects because they will work with a good range of supply voltages and they use very little power. However, if you are using them to design circuits and investigate logic gates please remember that all unused inputs MUST be connected to the power supply (either +Vs or 0V), this applies even if that part of the IC is not being used in the circuit.

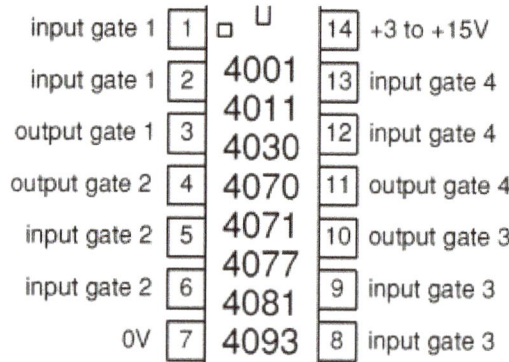

| | | | |
|---|---|---|---|
| input gate 1 | 1 | □ ⊔ | 14 +3 to +15V |
| input gate 1 | 2 | 4001 | 13 input gate 4 |
| output gate 1 | 3 | 4011 4030 | 12 input gate 4 |
| output gate 2 | 4 | 4070 | 11 output gate 4 |
| input gate 2 | 5 | 4071 | 10 output gate 3 |
| input gate 2 | 6 | 4077 4081 | 9 input gate 3 |
| 0V | 7 | 4093 | 8 input gate 3 |

## Not Gate

A NOT gate can only have one input and the output is the inverse of the input. A NOT gate is also called an inverter.

The output Q is true when the input A is NOT true: Q = NOT A

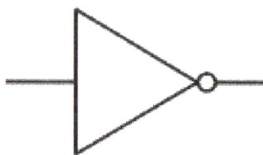

| Traditional symbol | IEC symbol |
|---|---|

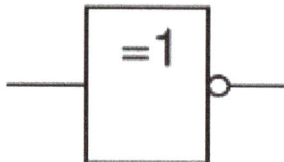

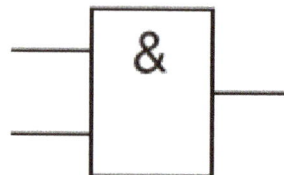

| Input A | Output Q |
|---|---|
| 0 | 1 |
| 1 | 0 |

## AND Gate

An AND gate can have two or more inputs, its output is true if all inputs are true. The output Q is true if input A AND input B are both true: Q = A AND B

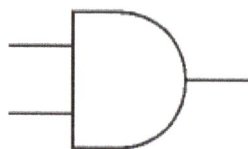

| Traditional symbol | IEC symbol |
|---|---|

| Input A | Input B | Output Q |
|---|---|---|
| 0 | 0 | 1 |

| 0 | 1 | 1 |
|---|---|---|
| 1 | 0 | 1 |
| 1 | 1 | 0 |

## NAND Gate

NAND = Not AND. This is an AND gate with the output inverted, as shown by the 'o' on the symbol output. A NAND gate can have two or more inputs, its output is true if NOT all inputs are true. The output Q is true if input A AND input B are NOT both true: Q = NOT (A AND B).

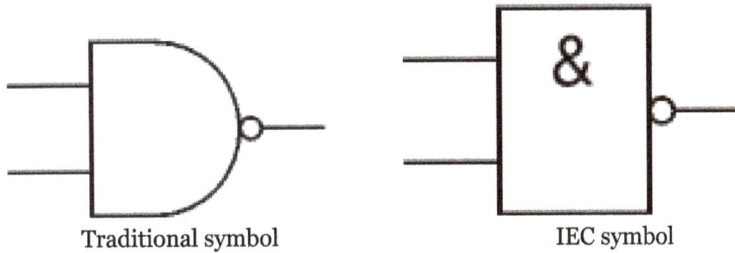

Traditional symbol                                    IEC symbol

| Input A | Input B | Output Q |
|---------|---------|----------|
| 0 | 0 | 1 |
| 0 | 1 | 1 |
| 1 | 0 | 1 |
| 1 | 1 | 0 |

## OR Gate

An OR gate can have two or more inputs, its output is true if at least one input is true. The output Q is true if input A OR input B is true (or both of them are true): Q = A OR B.

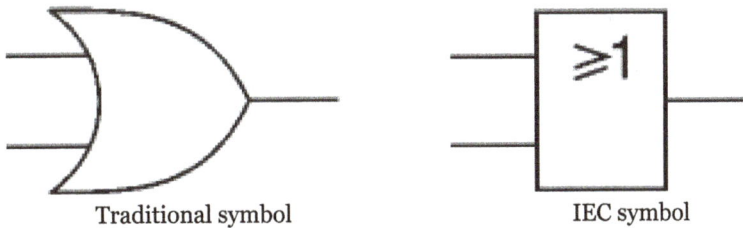

Traditional symbol                                    IEC symbol

| Input A | Input B | Output Q |
|---------|---------|----------|
| 0 | 0 | 0 |
| 0 | 1 | 1 |
| 1 | 0 | 1 |
| 1 | 1 | 1 |

## NOR Gate

NOR = Not OR. This is an OR gate with the output inverted, as shown by the 'o' on the symbol output. A NOR gate can have two or more inputs, its output is true if no inputs are true. The output Q is true if NOT inputs A OR B are true: Q = NOT (A OR B).

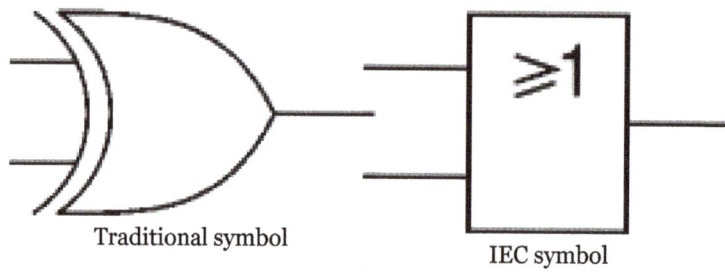

Traditional symbol

IEC symbol

| Input A | Input B | Output Q |
|---------|---------|----------|
| 0 | 0 | 1 |
| 0 | 1 | 0 |
| 1 | 0 | 0 |
| 1 | 1 | 0 |

## EX-OR Gate

EXclusive-OR. This is like an OR gate but excluding both inputs being true. The output is true if inputs A and B are DIFFERENT. EX-OR gates can only have 2 inputs. The output Q is true if either input A is true OR input B is true, but not when both of them are true: Q = (A AND NOT B) OR ( B AND NOT A).

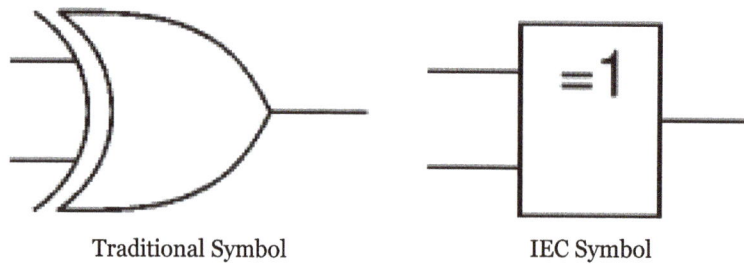

Traditional Symbol

IEC Symbol

| Input A | Input B | Output Q |
|---------|---------|----------|
| 0 | 0 | 0 |
| 0 | 1 | 1 |
| 1 | 0 | 1 |
| 1 | 1 | 0 |

## EX-NOR Gate

EXclusive-NOR. This is an EX-OR gate with the output inverted, as shown by the 'o' on the symbol output. EX-NOR gates can only have 2 inputs. The output Q is true if inputs A and B are the SAME(both true or both false): Q = (A AND B) OR (NOT A AND NOT B).

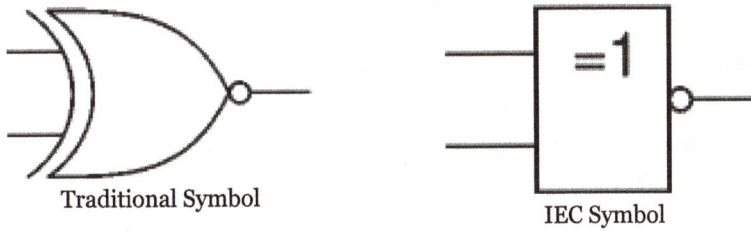

Traditional Symbol

IEC Symbol

| Input A | Input B | Output Q |
|---------|---------|----------|
| 0 | 0 | 1 |
| 0 | 1 | 0 |
| 1 | 0 | 0 |
| 1 | 1 | 1 |

## Combinations of Logic Gates

Logic gates can be combined to produce more complex functions.

For example to produce an output Q which is true only when input A is true and input B is false, we can combine a NOT gate and an AND gate as shown.

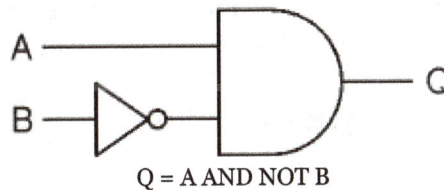

Q = A AND NOT B

## Working out The Function of A Combination of Gates

Truth tables can be used to work out the function of a combination of gates such as the system shown below:

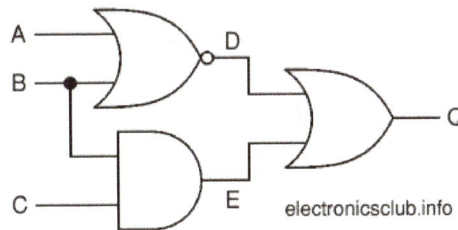

electronicsclub.info

Begin by creating a table showing all possible combinations of inputs (A, B and C in this example) with enough extra columns for each intermediate output (D and E in this example) as well as the final output (Q). Then work out all the intermediate output states, filling in the table as you do go. These intermediate outputs form the inputs to the next gate (or gates) so you can use these to work out the next output(s), in this example that is for the final output (Q).

D = NOT (A OR B)

E = B AND C

$$Q = D \text{ OR } E = (\text{NOT } (A \text{ OR } B)) \text{ OR } (B \text{ AND } C)$$

The truth table shows the intermediate outputs D and E as well as the final output Q.

| Inputs | | | Outputs | | |
|---|---|---|---|---|---|
| A | B | C | D | E | Q |
| 0 | 0 | 0 | 1 | 0 | 1 |
| 0 | 0 | 1 | 1 | 0 | 1 |
| 0 | 1 | 0 | 0 | 0 | 0 |
| 0 | 1 | 1 | 0 | 1 | 1 |
| 1 | 0 | 0 | 0 | 0 | 0 |
| 1 | 0 | 1 | 0 | 0 | 0 |
| 1 | 1 | 0 | 0 | 0 | 0 |
| 1 | 1 | 1 | 0 | 1 | 1 |

## Substituting One Type of Gate for another

Logic gates are available on ICs which usually contain several gates of the same type, for example four 2-input NAND gates or three 3-input NAND gates. This can be wasteful if only a few gates are required unless they are all the same type. To avoid using too many ICs you can reduce the number of gate inputs or substitute one type of gate for another.

## Reducing the Number of Inputs

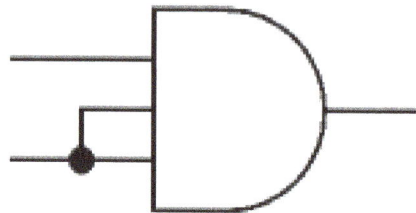

The number of inputs to a gate can be reduced by connecting two (or more) inputs together. The diagram shows a 3-input AND gate operating as a 2-input AND gate.

Making a NOT gate from a NAND or NOR gate

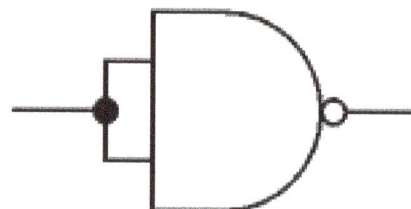

Reducing a NAND or NOR gate to just one input creates a NOT gate. The diagram shows this for a 2-input NAND gate.

## Any Gate Can be Built From Nand Or Nor Gates

As well as making a NOT gate, NAND or NOR gates can be combined to create any type of gate! This enables a circuit to be built from just one type of gate, either NAND or NOR. For example an AND gate is a NAND gate then a NOT gate (to undo the inverting function). Note that AND and OR gates cannot be used to create other gates because they lack the inverting (NOT) function.

To change the type of gate, such as changing OR to AND, you must do three things:

- Invert (NOT) each input.

- Change the gate type (OR to AND, or AND to OR)

- Invert (NOT) the output.

For example an OR gate can be built from NOTed inputs fed into a NAND (AND + NOT) gate.

## NAND gate equivalents

The arrangements below show how to use NAND gates to make NOT, AND, OR and NOR gates:

NOT made from one NAND gate:

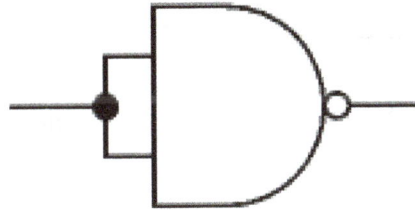

AND made from two NAND gates:

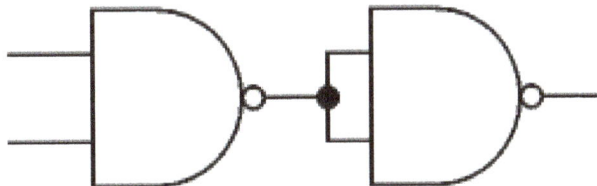

OR made from three NAND gates:

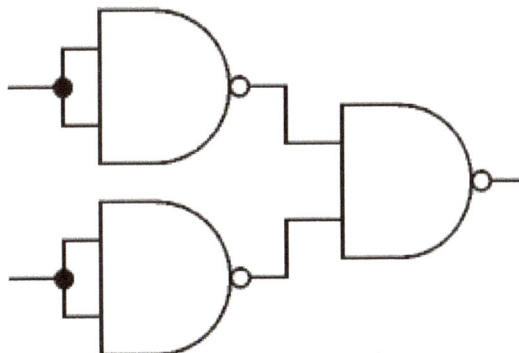

NOR made from four NAND gates:

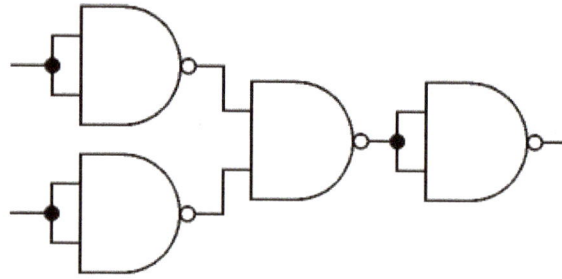

Substituting gates in an example logic system

This system has 3 different gates (NOR, AND and OR) so it requires three ICs, one for each type of gate.

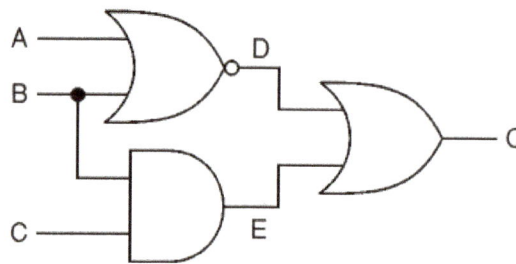

To re-design this system using NAND gates only begin by replacing each gate with its NAND gate equivalent, as shown below:

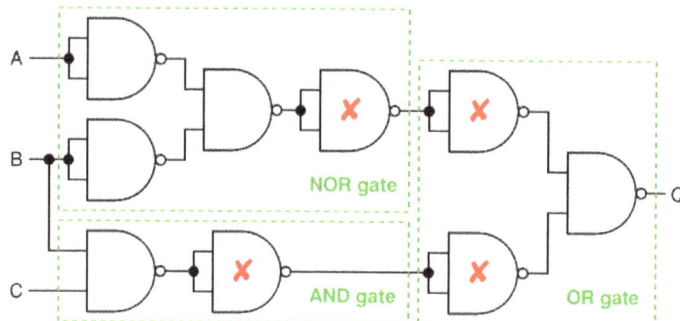

Then simplify the system by deleting adjacent pairs of NOT gates (marked X above). This can be done because the second NOT gate cancels the action of the first:

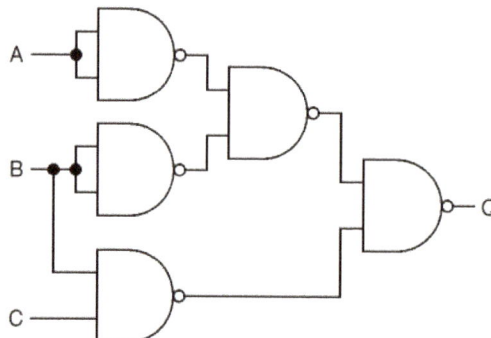

The final system has five NAND gates and requires two ICs (with four gates on each IC). This is better than the original system which required three ICs (one for each type of gate).

Substituting NAND (or NOR) gates does not always increase the number of gates, but when it does (as in this example) the increase is usually only one or two gates. The real benefit is reducing the number of ICs required by using just one type of gate.

## Combinational Logic

Combinatorial logic is a concept in which two or more input states define one or more output states, where the resulting state or states are related by defined rules that are independent of previous states. Each of the inputs and output(s) can attain either of two states: logic 0 (low) or logic 1 (high). A common example is a simple logic gate .

In combinatorial logic, the output is a function of the input at a specific time. An example is the AND gate. Suppose an AND gate has two inputs, X and Y, and one output, Z. The AND operation is symbolized by an asterisk (*). The operation of the two-input AND gate can be represented by the following truth table:

| X | Y | X * Y |
|---|---|-------|
|   |   |       |
|   | 1 |       |
| 1 |   |       |
| 1 | 1 | 1     |

The output of this logic gate is high if and only if both inputs are high. If either or both of the inputs are low, the output is low. If the states of the inputs change periodically, the output always obeys the rules depicted in the above truth table. The current state of the device is not affected by preceding states. This contrasts combinatorial logic with sequential logic , in which the current state of the device is affected by previous states.

## Combinational Logic Circuits

Unlike Sequential Logic Circuits whose outputs are dependent on both their present inputs and their previous output state giving them some form of Memory.

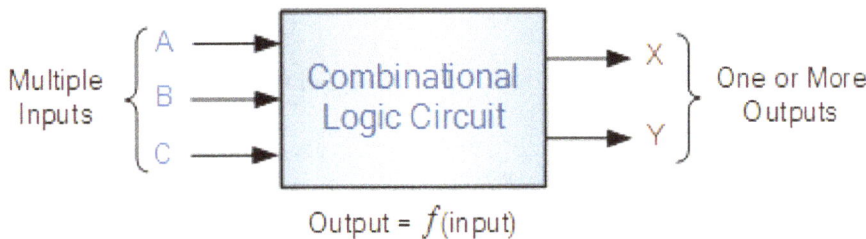

Output = $f$(input)

The outputs of Combinational Logic Circuits are only determined by the logical function of their current input state, logic "0" or logic "1", at any given instant in time.

The result is that combinational logic circuits have no feedback, and any changes to the signals being applied to their inputs will immediately have an effect at the output. In other words, in a Combinational Logic Circuit, the output is dependent at all times on the combination of its inputs. So if one of its inputs condition changes state, from 0-1 or 1-0, so too will the resulting output as by default combinational logic circuits have "no memory", "timing" or "feedback loops" within their design.

## Combinational Logic

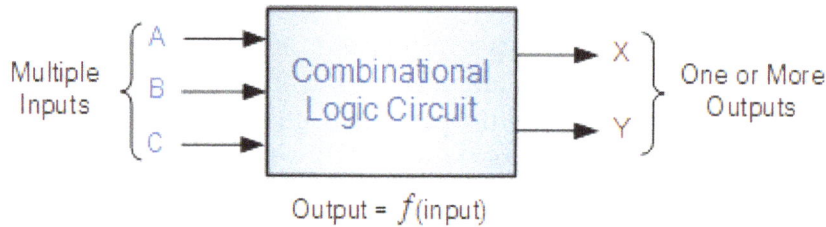

Multiple Inputs { A, B, C } → Combinational Logic Circuit → { X, Y } One or More Outputs

Output = $f$(input)

Combinational Logic Circuits are made up from basic logic NAND, NOR or NOT gates that are "combined" or connected together to produce more complicated switching circuits. These logic gates are the building blocks of combinational logic circuits. An example of a combinational circuit is a decoder, which converts the binary code data present at its input into a number of different output lines, one at a time producing an equivalent decimal code at its output.

Combinational logic circuits can be very simple or very complicated and any combinational circuit can be implemented with only NAND and NOR gates as these are classed as "universal" gates.

The three main ways of specifying the function of a combinational logic circuit are:

1. Boolean Algebra – This forms the algebraic expression showing the operation of the logic circuit for each input variable either True or False that result in a logic "1" output.

2. Truth Table – A truth table defines the function of a logic gate by providing a concise list that shows all the output states in tabular form for each possible combination of input variable that the gate could encounter.

3. Logic Diagram – This is a graphical representation of a logic circuit that shows the wiring and connections of each individual logic gate, represented by a specific graphical symbol, that implements the logic circuit.

All three of these logic circuit representations are shown below.

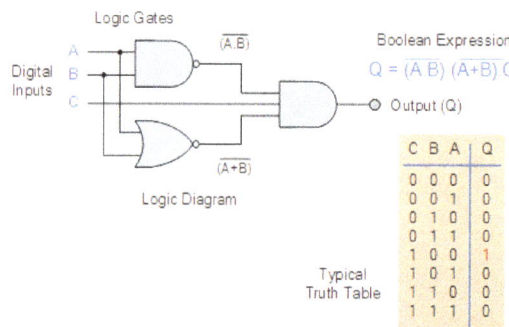

Logic Gates

Digital Inputs: A, B, C

$(\overline{A.B})$

$(A+B)$

Logic Diagram

Boolean Expression

$Q = (\overline{A.B})\,(A+B)\,C$

Output (Q)

Typical Truth Table

| C | B | A | Q |
|---|---|---|---|
| 0 | 0 | 0 | 0 |
| 0 | 0 | 1 | 0 |
| 0 | 1 | 0 | 0 |
| 0 | 1 | 1 | 0 |
| 1 | 0 | 0 | 1 |
| 1 | 0 | 1 | 0 |
| 1 | 1 | 0 | 0 |
| 1 | 1 | 1 | 0 |

As combinational logic circuits are made up from individual logic gates only, they can also be considered as "decision making circuits" and combinational logic is about combining logic gates together to process two or more signals in order to produce at least one output signal according to the logical function of each logic gate. Common combinational circuits made up from individual logic gates that carry out a desired application.

Include Multiplexers, De-multiplexers, Encoders, Decoders, Full and Half Adders etc.

Classification of Combinational Logic

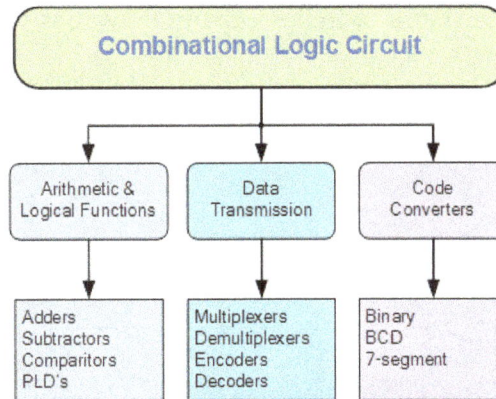

One of the most common uses of combinational logic is in Multiplexer and De-multiplexer type circuits. Here, multiple inputs or outputs are connected to a common signal line and logic gates are used to decode an address to select a single data input or output switch.

A multiplexer consist of two separate components, a logic decoder and some solid state switches, but before we can discuss multiplexers, decoders and de-multiplexers in more detail we first need to understand how these devices use these "solid state switches" in their design.

## Solid State Switches

Standard TTL logic devices made up from Transistors can only pass signal currents in one direction only making them "uni-directional" devices and poor imitations of conventional electro-mechanical switches or relays. However, some CMOS switching devices made up from FET's act as near perfect "bi-directional" switches making them ideal for use as solid state switches.

Solid state switches come in a variety of different types and ratings, and there are many different applications for using solid state switches. They can basically be sub-divided into 3 different main groups for switching applications and in this combinational logic section we will only look at the Analogue type of switch but the principal is the same for all types including digital.

## Solid State Switch Applications

- Analogue Switches – Used in Data Switching and Communications, Video and Audio Signal Switching, Instrumentation and Process Control Circuits etc.

- Digital Switches – High Speed Data Transmission, Switching and Signal Routing, Ethernet, LAN's, USB and Serial Transmissions etc.

- Power Switches – Power Supplies and General "Standby Power" Switching Applications, Switching of Larger Voltages and Currents etc.

## Analogue Bilateral Switches

Analogue or "Analog" switches are those types that are used to switch data or signal currents when they are in their "ON" state and block them when they are in their "OFF" state. The rapid switching between the "ON" and the "OFF" state is usually controlled by a digital signal applied to the control gate of the switch. An ideal analogue switch has zero resistance when "ON" (or closed), and infinite resistance when "OFF" (or open) and switches with RON values of less than 1Ω are commonly available.

## Solid State Analogue Switch

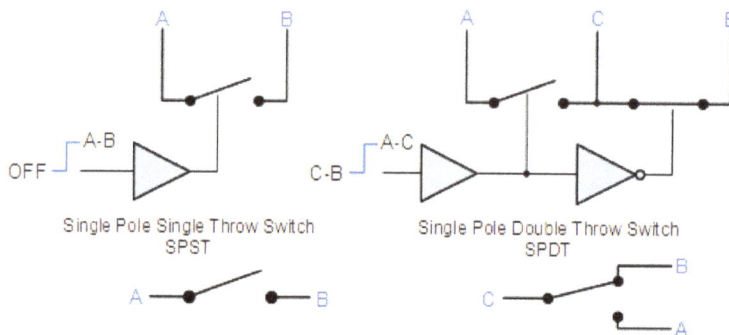

By connecting an N-channel MOSFET in parallel with a P-channel MOSFET allows signals to pass in either direction making it a Bi-directional switch and as to whether the N-channel or the P-channel device carries more signal current will depend upon the ratio between the input to the output voltage. The two MOSFET's are switched "ON" or "OFF" by two internal non-inverting and inverting amplifiers.

## Contact Types

Just like mechanical switches, analogue switches come in a variety of forms or contact types, depending on the number of "poles" and "throws" they offer. Thus, terms such as "SPST" (single-pole single throw) and "SPDT" (single-pole double-throw) also apply to solid state analogue switches with "make-before-break" and "break-before-make" configurations available.

## Analogue Switch Types

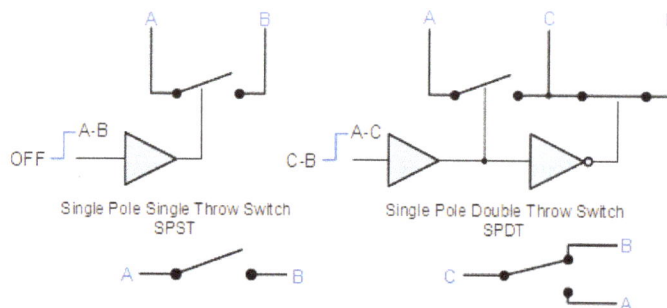

Individual analogue switches can be grouped together into standard IC packages to form devices with multiple switching configurations of SPST (single-pole single-throw) and SPDT (single-pole double-throw) as well as multi-channel multiplexers.

The most common and simplest analogue switch in a single IC package is the 74HC4066 which has 4 independent bi-directional "ON/OFF" Switches within a single package but the most widely used variants of the CMOS analogue switch are those described as "Multi-way Bilateral Switches" otherwise known as the "Multiplexer" and "De-multiplexer" IC´s.

## Half Adder

Half adder is a combinational logic circuit with two inputs and two outputs. The half adder circuit is designed to add two single bit binary numbers A and B. It is the basic building block for addition of two single bit numbers. This circuit has two outputs carry and sum.

## Block Diagram

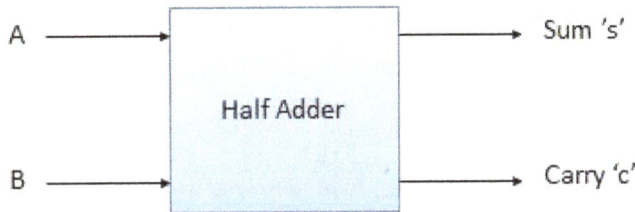

## Truth Table

| Inputs | | Output | |
|---|---|---|---|
| A | B | S | C |
| 0 | 0 | 0 | 0 |
| 0 | 1 | 1 | 0 |
| 1 | 0 | 1 | 0 |
| 1 | 1 | 0 | 1 |

## Circuit Diagram

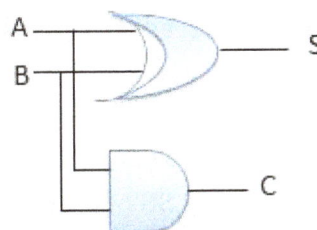

## Full Adder

Full adder is developed to overcome the drawback of Half Adder circuit. It can add two one-bit numbers A and B, and carry c. The full adder is a three input and two output combinational circuit.

## Block Diagram

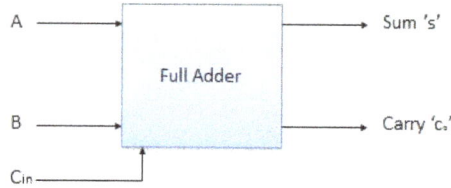

## Truth Table

| Inputs | | | Output | |
|---|---|---|---|---|
| A | B | Cin | S | Co |
| 0 | 0 | 0 | 0 | 0 |
| 0 | 0 | 1 | 1 | 0 |
| 0 | 1 | 0 | 1 | 0 |
| 0 | 1 | 1 | 0 | 1 |
| 1 | 0 | 0 | 1 | 0 |
| 1 | 0 | 1 | 0 | 1 |
| 1 | 1 | 0 | 0 | 1 |
| 1 | 1 | 1 | 1 | 1 |

## Circuit Diagram

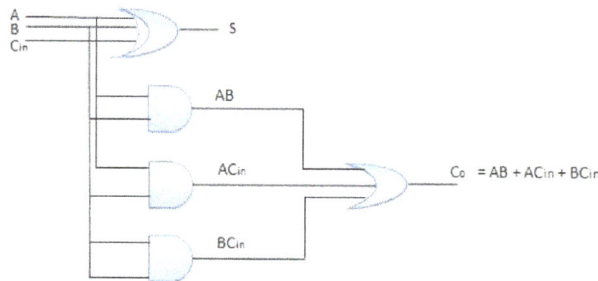

## N-Bit Parallel Adder

The Full Adder is capable of adding only two single digit binary number along with a carry input. But in practical we need to add binary numbers which are much longer than just one bit. To add two n-bit binary numbers we need to use the n-bit parallel adder. It uses a number of full adders in cascade. The carry output of the previous full adder is connected to carry input of the next full adder.

## 4 Bit Parallel Adder

In the block diagram, A0 and B0 represent the LSB of the four bit words A and B. Hence Full Adder-o is the lowest stage. Hence its Cin has been permanently made 0. The rest of the connections are exactly same as those of n-bit parallel adder is shown in figure The four bit parallel adder is a very common logic circuit.

## Block Diagram

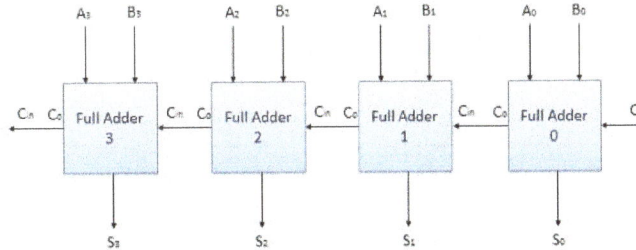

## N-Bit Parallel Subtractor

The subtraction can be carried out by taking the 1's or 2's complement of the number to be subtracted. For example we can perform the subtraction (A-B) by adding either 1's or 2's complement of B to A. That means we can use a binary adder to perform the binary subtraction.

## 4 Bit Parallel Subtractor

The number to be subtracted (B) is first passed through inverters to obtain its 1's complement. The 4-bit adder then adds A and 2's complement of B to produce the subtraction. S3 S2 S1 S0 represents the result of binary subtraction (A-B) and carry output Cout represents the polarity of the result. If A > B then Cout = 0 and the result of binary form (A-B) then Cout = 1 and the result is in the 2's complement form.

## Block Diagram

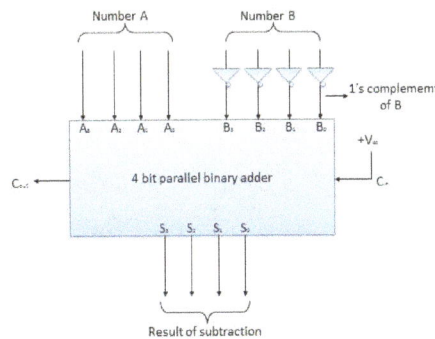

## Half Subtractors

Half subtractor is a combination circuit with two inputs and two outputs (difference and borrow). It produces the difference between the two binary bits at the input and also produces an output (Borrow) to indicate if a 1 has been borrowed. In the subtraction (A-B), A is called as Minuend bit and B is called as Subtrahend bit.

## Truth Table

| Inputs | | Output | |
|---|---|---|---|
| A | B | (A – B) | Borrow |
| 0 | 0 | 0 | 0 |
| 0 | 1 | 1 | 1 |
| 1 | 0 | 1 | 0 |
| 1 | 1 | 0 | 0 |

## Circuit Diagram

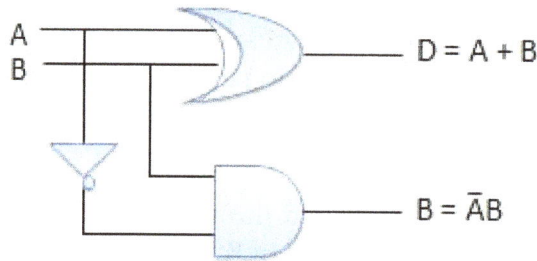

$D = A + B$

$B = \bar{A}B$

## Full Subtractors

The disadvantage of a half subtractor is overcome by full subtractor. The full subtractor is a combinational circuit with three inputs A,B,C and two output D and C'. A is the 'minuend', B is 'subtrahend', C is the 'borrow' produced by the previous stage, D is the difference output and C' is the borrow output.

## Truth Table

| Inputs | | | Output | |
|---|---|---|---|---|
| A | B | C | (A-B-C) | C' |
| 0 | 0 | 0 | 0 | 0 |
| 0 | 0 | 1 | 1 | 1 |
| 0 | 1 | 0 | 1 | 1 |
| 0 | 1 | 1 | 0 | 1 |
| 1 | 0 | 0 | 1 | 0 |
| 1 | 0 | 1 | 0 | 0 |
| 1 | 1 | 0 | 0 | 0 |
| 1 | 1 | 1 | 1 | 1 |

## Circuit Diagram

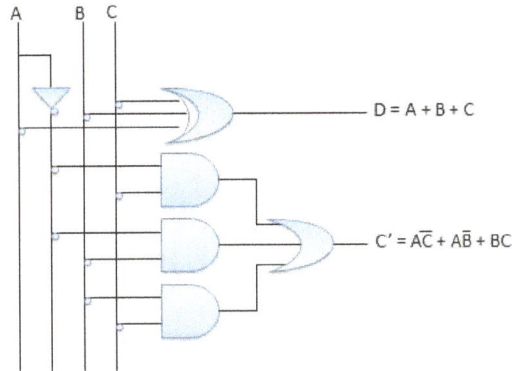

$D = A + B + C$

$C' = A\overline{C} + A\overline{B} + BC$

## Multiplexer

Multiplexer is a special type of combinational circuit. There are n-data inputs, one output and m select inputs with 2m = n. It is a digital circuit which selects one of the n data inputs and routes it to the output. The selection of one of the n inputs is done by the selected inputs. Depending on the digital code applied at the selected inputs, one out of n data sources is selected and transmitted to the single output Y. E is called the strobe or enable input which is useful for the cascading. It is generally an active low terminal that means it will perform the required operation when it is low.

## Block Diagram

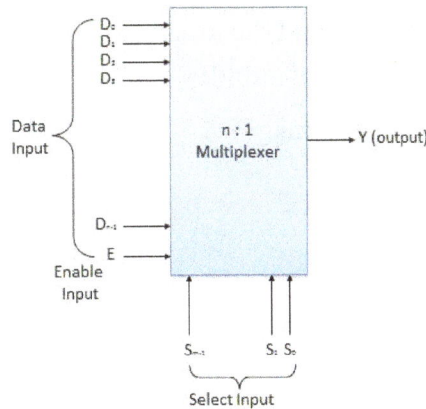

Multiplexers come in multiple variations

- 2 : 1 multiplexer
- 4 : 1 multiplexer
- 16 : 1 multiplexer
- 32 : 1 multiplexer

## Block Diagram

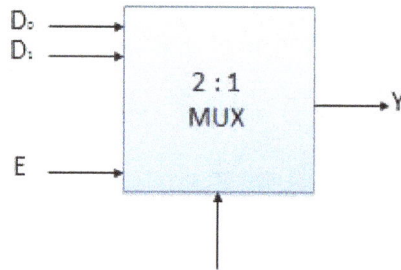

## Truth Table

| Enable | Select | Output |
|--------|--------|--------|
| E | S | Y |
| 0 | x | 0 |
| 1 | 0 | $D_0$ |
| 1 | 1 | $D_1$ |

x = Don't care

## Demultiplexers

A demultiplexer performs the reverse operation of a multiplexer i.e. it receives one input and distributes it over several outputs. It has only one input, n outputs, m select input. At a time only one output line is selected by the select lines and the input is transmitted to the selected output line. A de-multiplexer is equivalent to a single pole multiple way switch as shown in figure.

Demultiplexers comes in multiple variations.

- 1 : 2 demultiplexer
- 1 : 4 demultiplexer
- 1 : 16 demultiplexer
- 1 : 32 demultiplexer

## Block Diagram

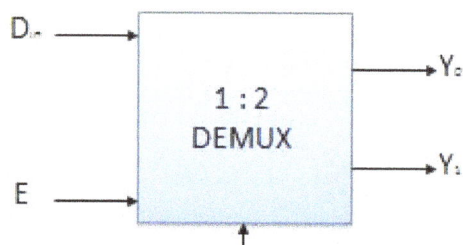

## Truth Table

| Enable | Select | Output | |
|---|---|---|---|
| E | S | Y0 | Y1 |
| 0 | x | 0 | 0 |
| 1 | 0 | 0 | $D_{in}$ |
| 1 | 1 | $D_{in}$ | 0 |

x = Don't care

## Decoder

A decoder is a combinational circuit. It has n input and to a maximum m = 2n outputs. Decoder is identical to a demultiplexer without any data input. It performs operations which are exactly opposite to those of an encoder.

## Block Diagram

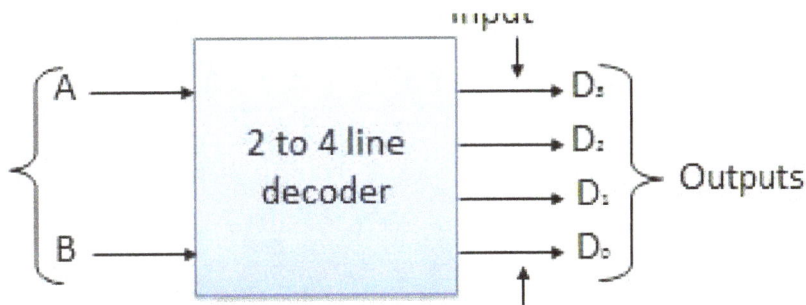

Examples of Decoders are following.

- Code converters
- BCD to seven segment decoders
- Nixie tube decoders
- Relay actuator

## 2 to 4 Line Decoder

The block diagram of 2 to 4 line decoder is shown in the figure A and B are the two inputs where D through D are the four outputs. Truth table explains the operations of a decoder. It shows that each output is 1 for only a specific combination of inputs.

## Block diagram

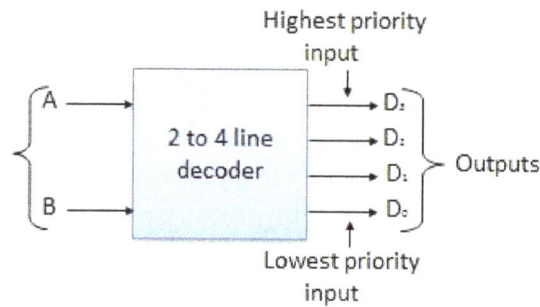

## Truth Table

| Inputs | | Output | | | |
|---|---|---|---|---|---|
| A | B | $D_0$ | $D_1$ | $D_2$ | $D_3$ |
| 0 | 0 | 1 | 0 | 0 | 0 |
| 0 | 1 | 0 | 1 | 0 | 0 |
| 0 | 1 | 0 | 0 | 1 | 0 |
| 1 | 1 | 0 | 0 | 0 | 1 |

## Logic Circuit

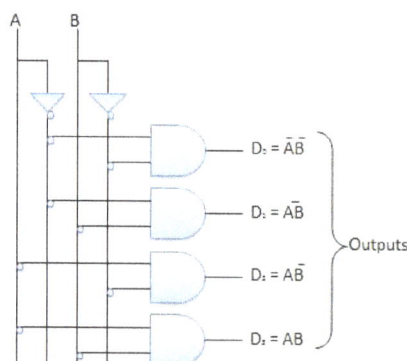

## Encoder

Encoder is a combinational circuit which is designed to perform the inverse operation of the decoder. An encoder has n number of input lines and m number of output lines. An encoder produces an m bit binary code corresponding to the digital input number. The encoder accepts an n input digital word and converts it into an m bit another digital word.

## Block Diagram

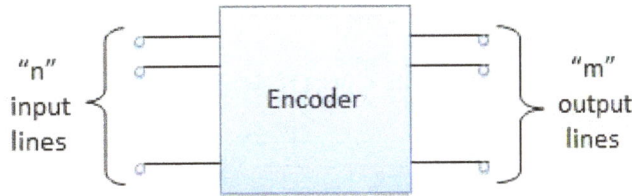

Examples of Encoders are following.

- Priority encoders

- Decimal to BCD encoder

- Octal to binary encoder

- Hexadecimal to binary encoder

## Priority Encoder

This is a special type of encoder. Priority is given to the input lines. If two or more input line are 1 at the same time, then the input line with highest priority will be considered. There are four inputs $D_0$, $D_1$, $D_2$, $D_3$ and two output $Y_0$, $Y_1$. Out of the four input $D_3$ has the highest priority and $D_0$ has the lowest priority. That means if $D_3 = 1$ then $Y_1 Y_1 = 11$ irrespective of the other inputs. Similarly if $D_3 = 0$ and $D_2 = 1$ then $Y_1 Y_0 = 10$ irrespective of the other inputs.

## Block Diagram

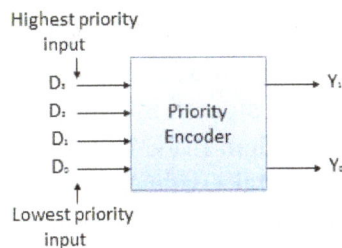

## Truth Table

| Highest | Inputs | | Lowest | Outputs | |
|---|---|---|---|---|---|
| $D_3$ | $D_2$ | $D_1$ | $D_0$ | $Y_0$ | $Y_1$ |
| 0 | 0 | 0 | 0 | x | x |
| 0 | 0 | 0 | 1 | 0 | 0 |
| 0 | 0 | 1 | x | 0 | 1 |
| 0 | 1 | x | x | 1 | 0 |
| 1 | x | x | x | 1 | 1 |

## Logic Circuit

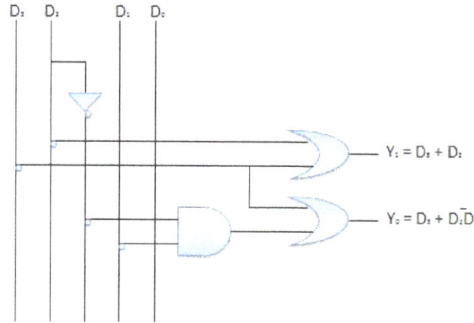

# CMOS

The term CMOS stands for "Complementary Metal Oxide Semiconductor". CMOS technology is one of the most popular technology in the computer chip design industry and broadly used today to form integrated circuits in numerous and varied applications. Today's computer memories, CPUs and cell phones make use of this technology due to several key advantages. This technology makes use of both P channel and N channel semiconductor devices.

One of the most popular MOSFET technologies available today is the Complementary MOS or CMOS technology. This is the dominant semiconductor technology for microprocessors, microcontroller chips, memories like RAM, ROM, EEPROM and application specific integrated circuits (ASICs).

The main advantage of CMOS over NMOS and BIPOLAR technology is the much smaller power dissipation. Unlike NMOS or BIPOLAR circuits, a Complementary MOS circuit has almost no static power dissipation. Power is only dissipated in case the circuit actually switches. This allows integrating more CMOS gates on an IC than in NMOS or bipolar technology, resulting in much better performance. Complementary Metal Oxide Semiconductor transistor consists P-channel MOS (PMOS) and N-channel MOS (NMOS).

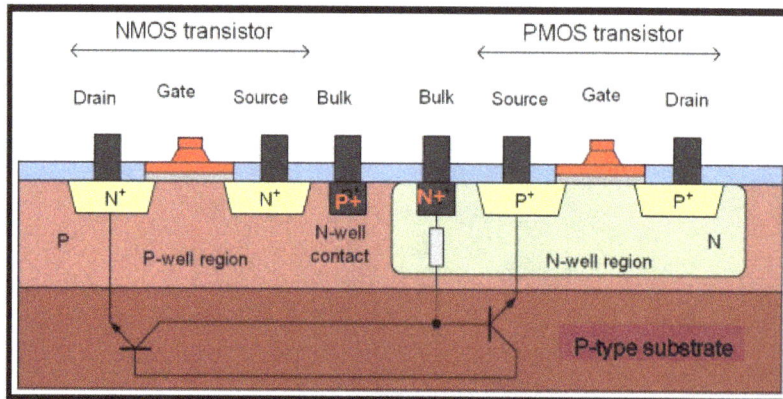

CMOS Transistor

## NMOS

NMOS is built on a p-type substrate with n-type source and drain diffused on it. In NMOS, the majority carriers are electrons. When a high voltage is applied to the gate, the NMOS will conduct. Similarly, when a low voltage is applied to the gate, NMOS will not conduct. NMOS are considered to be faster than PMOS, since the carriers in NMOS, which are electrons, travel twice as fast as the holes.

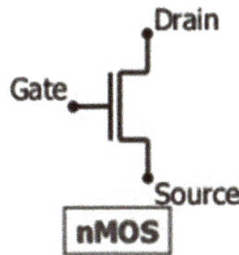

NMOS Transistor

## PMOS

P- channel MOSFET consists P-type Source and Drain diffused on an N-type substrate. Majority carriers are holes. When a high voltage is applied to the gate, the PMOS will not conduct. When a low voltage is applied to the gate, the PMOS will conduct. The PMOS devices are more immune to noise than NMOS devices.

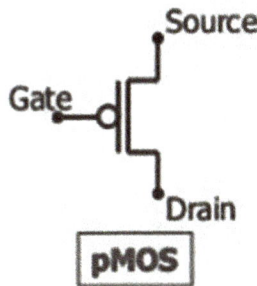

PMOS Transistor

## CMOS Working Principle

In CMOS technology, both N-type and P-type transistors are used to design logic functions. The same signal which turns ON a transistor of one type is used to turn OFF a transistor of the other type. This characteristic allows the design of logic devices using only simple switches, without the need for a pull-up resistor.

In CMOS logic gates a collection of n-type MOSFETs is arranged in a pull-down network between the output and the low voltage power supply rail (Vss or quite often ground). Instead of the load resistor of NMOS logic gates, CMOS logic gates have a collection of p-type MOSFETs in a pull-up network between the output and the higher-voltage rail (often named Vdd).

Thus, if both a p-type and n-type transistor have their gates connected to the same input, the p-type MOSFET will be ON when the n-type MOSFET is OFF, and vice-versa. The networks are arranged such that one is ON and the other OFF for any input pattern as shown in the figure below.

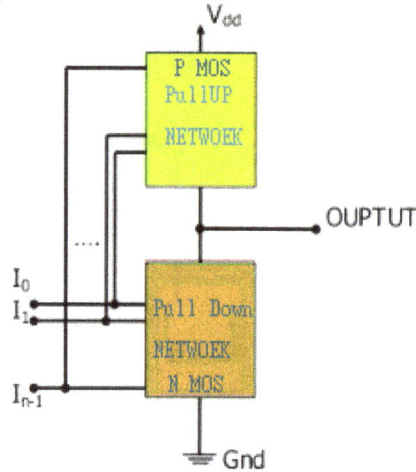

CMOS Logic Gate using Pull-Up and Pull-Down Networks

CMOS offers relatively high speed, low power dissipation, high noise margins in both states, and will operate over a wide range of source and input voltages (provided the source voltage is fixed).

## CMOS Inverter

The inverter circuit as shown in the figure below. It consists of PMOS and NMOS FET. The input A serves as the gate voltage for both transistors.

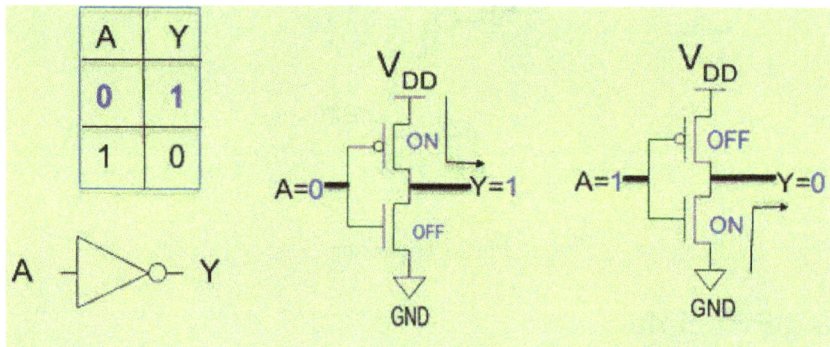

CMOS Inverter

The NMOS transistor has an input from Vss (ground) and PMOS transistor has an input from Vdd. The terminal Y is output. When a high voltage (~ Vdd) is given at input terminal (A) of the inverter, the PMOS becomes open circuit and NMOS switched OFF so the output will be pulled down to Vss.

When a low-level voltage (<Vdd, ~0v) applied to the inverter, the NMOS switched OFF and PMOS switched ON. So the output becomes Vdd or the circuit is pulled up to Vdd.

| INPUT | LOGIC INPUT | OUTPUT | LOGIC OUTPUT |
|---|---|---|---|
| 0 v | 0 | $V_{dd}$ | 1 |
| $V_{dd}$ | 1 | 0 v | 0 |

## CMOS NAND Gate

The below figure shows a 2-input Complementary MOS NAND gate. It consists of two series NMOS transistors between Y and Ground and two parallel PMOS transistors between Y and VDD.

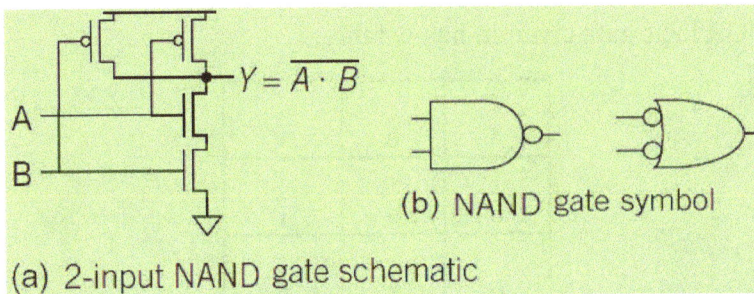

(a) 2-input NAND gate schematic

(b) NAND gate symbol

## CMOS NAND Gate

If either input A or B is logic 0, at least one of the NMOS transistors will be OFF, breaking the path from Y to Ground. But at least one of the pMOS transistors will be ON, creating a path from Y to VDD.

Hence, the output Y will be high. If both inputs are high, both of the nMOS transistors will be ON and both of the pMOS transistors will be OFF. Hence, the output will be logic low. The truth table of NAND logic gate given in table below.

| A | B | Pull-Down Network | Pull-up Network | OUTPUT Y |
|---|---|---|---|---|
| 0 | 0 | OFF | ON | 1 |
| 0 | 1 | OFF | ON | 1 |
| 1 | 0 | OFF | ON | 1 |
| 1 | 1 | ON | OFF | 0 |

## CMOS NOR Gate

A 2-input NOR gate is shown in the figure below. The NMOS transistors are in parallel to pull the output low when either input is high. The PMOS transistors are in series to pull the output high when both inputs are low, as given in below table. The output is never left floating.

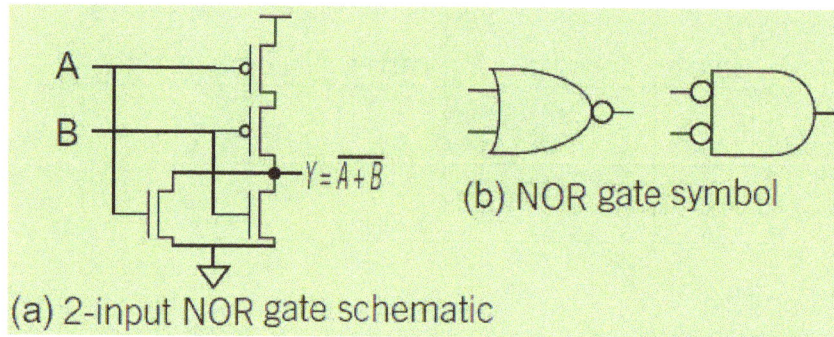

(b) NOR gate symbol

(a) 2-input NOR gate schematic

## Complementary MOS NOR Gate

The truth table of NOR logic gate given in below table.

| A | B | Y |
|---|---|---|
| 0 | 0 | 1 |
| 0 | 1 | 0 |
| 1 | 0 | 0 |
| 1 | 1 | 0 |

## CMOS Applications

Complementary MOS processes were widely implemented and have fundamentally replaced NMOS and bipolar processes for nearly all digital logic applications. The CMOS technology has been used for the following digital IC designs.

- Computer memories, CPUs

- Microprocessor designs

- Flash memory chip designing

- Used to design application specific integrated circuits (ASICs)

## CMOS Gate Circuitry

## Field-effect Transistors

Field-effect transistors, particularly the insulated-gate variety, may be used in the design of gate circuits. Being voltage-controlled rather than current-controlled devices, IGFETs tend to allow very simple circuit designs. Take for instance, the following inverter circuit built using P- and N-channel IGFETs:

Inverter circuit using IGFETs

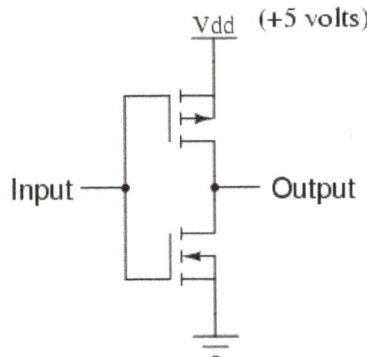

Notice the "Vdd" label on the positive power supply terminal. This label follows the same convention as "Vcc" in TTL circuits: it stands for the constant voltage applied to the drain of a field effect transistor, in reference to ground

## Field Effect Transistors in Gate Circuits

Let's connect this gate circuit to a power source and input switch, and examine its operation. Please note that these IGFET transistors are E-type (Enhancement-mode), and so are normally-off devices. It takes an applied voltage between gate and drain (actually, between gate and substrate) of the correct polarity to bias them on.

The upper transistor is a P-channel IGFET. When the channel (substrate) is made more positive than the gate (gate negative in reference to the substrate), the channel is enhanced and current is allowed between source and drain. So, in the above illustration, the top transistor is turned on.

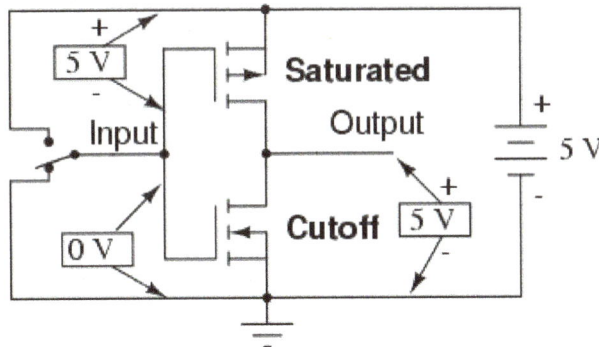

Input = "low" (0)

Output = "high" (1)

The lower transistor, having zero voltage between gate and substrate (source), is in its normal mode: off. Thus, the action of these two transistors are such that the output terminal of the gate circuit has a solid connection to Vdd and a very high resistance connection to ground. This makes the output "high" (1) for the "low" (0) state of the input.

Next, we'll move the input switch to its other position and see what happens:

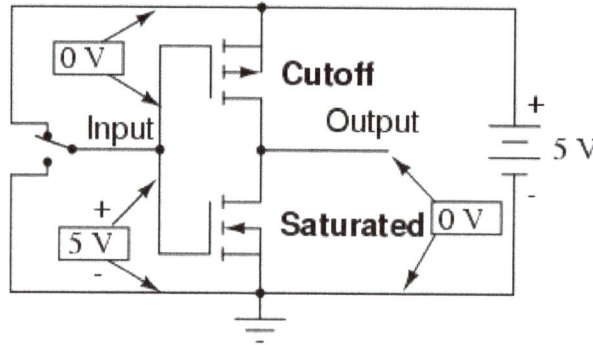

Input = "high" (1)

Output = "low" (0)

Now the lower transistor (N-channel) is saturated because it has sufficient voltage of the correct polarity applied between gate and substrate (channel) to turn it on (positive on gate, negative on the channel). The upper transistor, having zero voltage applied between its gate and substrate, is in its normal mode: off. Thus, the output of this gate circuit is now "low" (0). Clearly, this circuit exhibits the behavior of an inverter, or NOT gate.

## Complementary Metal Oxide Semiconductors

Using field-effect transistors instead of bipolar transistors has greatly simplified the design of the inverter gate. Note that the output of this gate never floats as is the case with the simplest TTL circuit: it has a natural "totem-pole" configuration, capable of both sourcing and sinking load current. Key to this gate circuit's elegant design is the complementary use of both P- and N-channel IGFETs. Since IGFETs are more commonly known as MOSFETs (Metal-Oxide-Semiconductor Field Effect Transistor), and this circuit uses both P- and N-channel transistors together, the general classification given to gate circuits like this one is CMOS: Complementary Metal Oxide Semiconductor.

## CMOS Gates: Challenges and Solutions

CMOS circuits aren't plagued by the inherent nonlinearities of the field-effect transistors, because as digital circuits their transistors always operate in either the saturated or cutoff modes and never in the active mode. Their inputs are, however, sensitive to high voltages generated by electrostatic (static electricity) sources, and may even be activated into "high" (1) or "low" (0) states by spurious voltage sources if left floating. For this reason, it is inadvisable to allow a CMOS logic gate input to float under any circumstances. Please note that this is very different from the behavior of a TTL gate where a floating input was safely interpreted as a "high" (1) logic level.

This may cause a problem if the input to a CMOS logic gate is driven by a single-throw switch, where one state has the input solidly connected to either Vdd or ground and the other state has the input floating (not connected to anything):

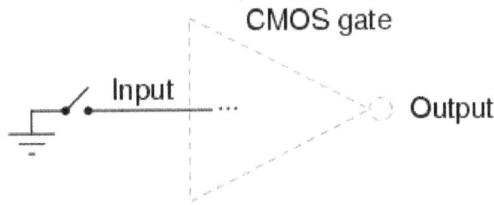

CMOS gate

Input

Output

*When switch is closed, the gate sees a definite "low" (0) input. However, when switch is open, the input logic level will be uncertain because it's floating.*

Also, this problem arises if a CMOS gate input is being driven by an open-collector TTL gate. Because such a TTL gate's output floats when it goes "high" (1), the CMOS gate input will be left in an uncertain state:

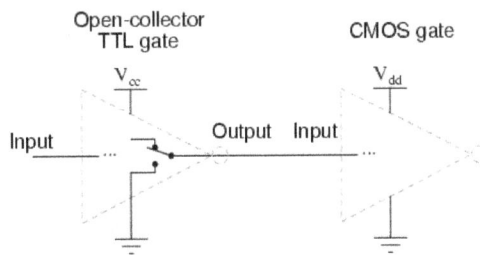

Open-collector TTL gate

CMOS gate

$V_{cc}$

$V_{dd}$

Input

Output   Input

*When the open-collector TTL gate's output is "high" (1), the CMOS gate's input will be left floating and in an uncertain logic state.*

Fortunately, there is an easy solution to this dilemma, one that is used frequently in CMOS logic circuitry. Whenever a single-throw switch (or any other sort of gate output incapable of both sourcing and sinking current) is being used to drive a CMOS input, a resistor connected to either Vdd or ground may be used to provide a stable logic level for the state in which the driving device's output is floating. This resistor's value is not critical: 10 kΩ is usually sufficient. When used to provide a "high" (1) logic level in the event of a floating signal source, this resistor is known as a pull-up resistor:

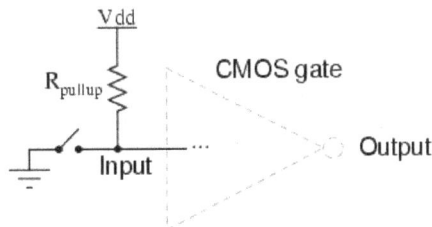

$V_{dd}$

$R_{pullup}$

CMOS gate

Input

Output

*When switch is closed, the gate sees a definite "low" (0) input. When the switch is open, $R_{pullup}$ will provide the connection to Vdd needed to secure a reliable "high" logic level for the CMOS gate input.*

When such a resistor is used to provide a "low" (0) logic level in the event of a floating signal source, it is known as a pull-down resistor. Again, the value for a pull-down resistor is not critical:

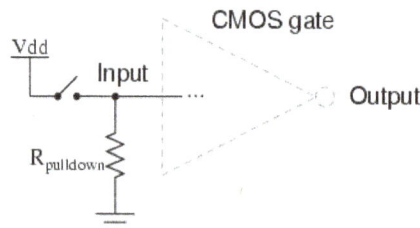

When switch is closed, the gate sees a definite "high" (1) input. When the switch is open, $R_{pulldown}$ will provide the connection to ground needed to secure a reliable "low" logic level for the CMOS gate input.

Because open-collector TTL outputs always sink, never source, current, pullup resistors are necessary when interfacing such an output to a CMOS gate input:

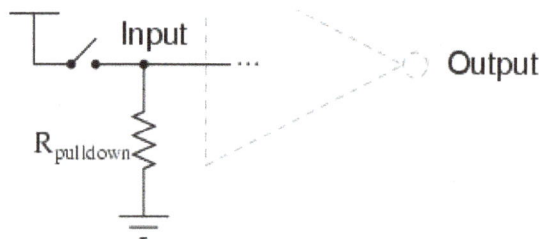

When switch is closed, the gate sees a definite "high" (1) input. When the switch is open, $R_{pulldown}$ will provide the connection

Although the CMOS gates used in the preceding examples were all inverters (single-input), the same principle of pull-up and pull-down resistors applies to multiple-input CMOS gates. Of course, a separate pull-up or pull-down resistor will be required for each gate input:

Pullup resistors for a 3-input
CMOS AND gate

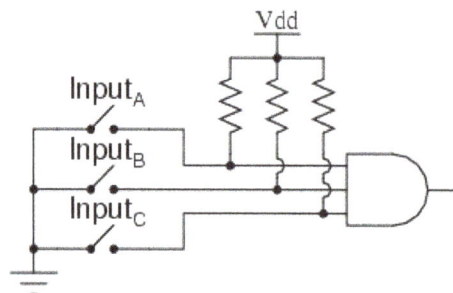

This brings us to the next question: how do we design multiple-input CMOS gates such as AND, NAND, OR, and NOR? Not surprisingly, the answer(s) to this question reveal simplicity of design much like that of the CMOS inverter over its TTL equivalent.

## CMOS NAND Gates

For example, here is the schematic diagram for a CMOS NAND gate:

CMOS NAND gate

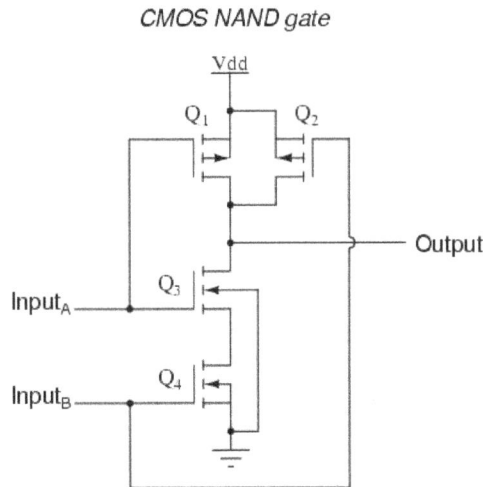

Notice how transistors Q1 and Q3 resemble the series-connected complementary pair from the inverter circuit. Both are controlled by the same input signal (input A), the upper transistor turning off and the lower transistor turning on when the input is "high" (1), and vice versa. Notice also how transistors Q2 and Q4 are similarly controlled by the same input signal (input B), and how they will also exhibit the same on/off behavior for the same input logic levels. The upper transistors of both pairs (Q1 and Q2) have their source and drain terminals paralleled, while the lower transistors (Q3 and Q4) are series-connected. What this means is that the output will go "high" (1) if either top transistor saturates, and will go "low" (0) only if both lower transistors saturate. The following sequence of illustrations shows the behavior of this NAND gate for all four possibilities of input logic levels (00, 01, 10, and 11):

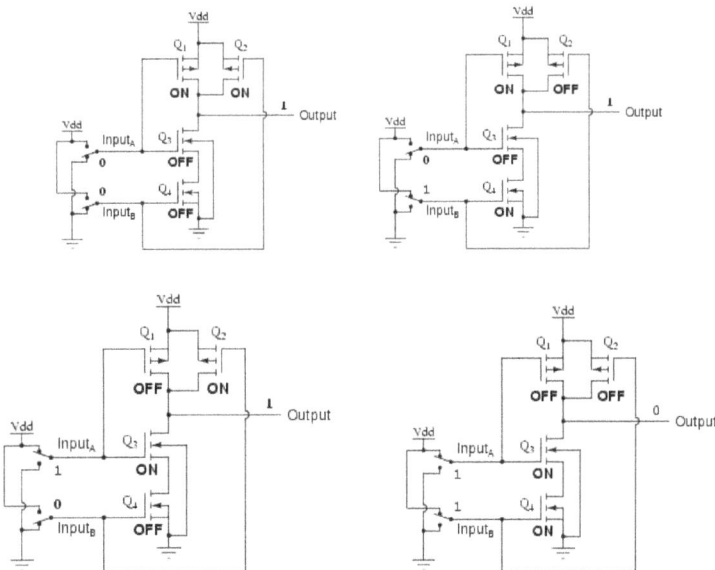

As with the TTL NAND gate, the CMOS NAND gate circuit may be used as the starting point for the creation of an AND gate. All that needs to be added is another stage of transistors to invert the output signal:

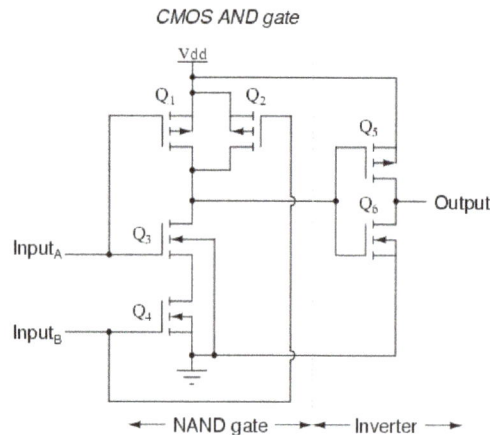

CMOS AND gate

## CMOS NOR Gates

A CMOS NOR gate circuit uses four MOSFETs just like the NAND gate, except that its transistors are differently arranged. Instead of two paralleled sourcing (upper) transistors connected to Vdd and two series-connected sinking (lower) transistors connected to ground, the NOR gate uses two series-connected sourcing transistors and two parallel-connected sinking transistors like this:

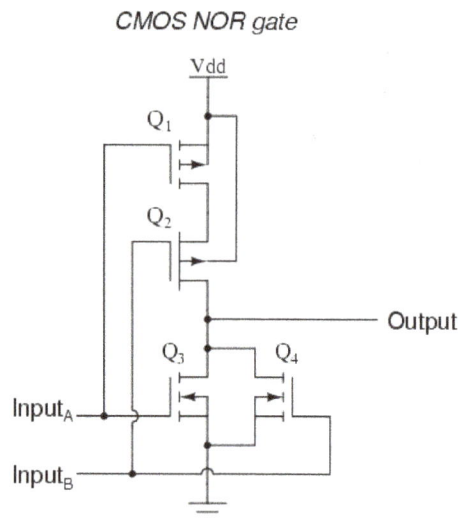

CMOS NOR gate

As with the NAND gate, transistors Q1 and Q3 work as a complementary pair, as do transistors Q2 and Q4. Each pair is controlled by a single input signal. If either input A or input B are "high" (1), at least one of the lower transistors (Q3 or Q4) will be saturated, thus making the output "low" (0). Only in the event of both inputs being "low" (0) will both lower transistors be in cutoff mode and both upper transistors be saturated, the conditions necessary for the output to go "high" (1). This behavior, of course, defines the NOR logic function.

## CMOS OR Gates

The OR function may be built up from the basic NOR gate with the addition of an inverter stage on the output:

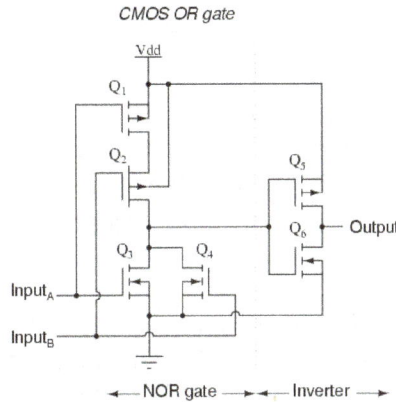
CMOS OR gate

# Logic Synthesis

Logic synthesis is the process of converting a high-level description of design into an optimized gate-level representation. Logic synthesis uses a standard cell library which have simple cells, such as basic logic gates like and, or, and nor, or macro cells, such as adder, muxes, memory, and flip-flops. Standard cells put together are called technology library. Normally the technology library is known by the transistor size (0.18u, 90nm).

A circuit description is written in Hardware Description Language (HDL) such as Verilog. The designer should first understand the architectural description. Then he should consider design constraints such as timing, area, testability, and power.

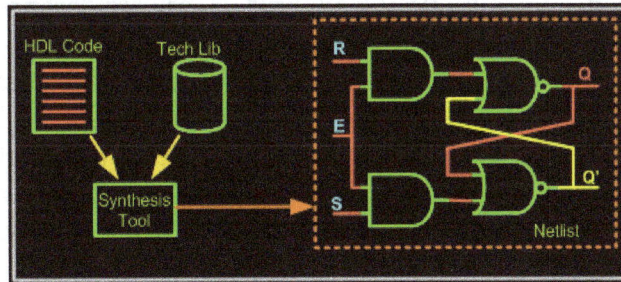

As you must have experienced in college, everything (all the digital circuits) is designed manually. Draw K-maps, optimize the logic, draw the schematic. This is how engineers used to design digital logic circuits in early days. Well this works fine as long as the design is a few hundred gates.

## Impact of Hdl and Logic Synthesis

High-level design is less prone to human error because designs are described at a higher level of abstraction. High-level design is done without significant concern about design constraints. Conversion from high-level design to gates is done by synthesis tools, using various algorithms to optimize the design as a whole. This removes the problem with varied designer styles for the different blocks in the design and suboptimal designs. Logic synthesis tools allow technology independent design. Design reuse is possible for technology-independent descriptions.

# High-level Synthesis

The growing capabilities of silicon technology and the increasing complexity of applications in recent decades have forced design methodologies and tools to move to higher abstraction levels. Raising the abstraction levels and accelerating automation of both the synthesis and the verification processes have for this reason always been key factors in the evolution of the design process, which in turn has allowed designers to explore the design space efficiently and rapidly

In the software domain, for example, machine code (binary sequence) was once the only language that could be used to program a computer. In the 1950s, the concept of assembly language (and assembler) was introduced. Finally, high-level languages (HLLs) and associated compilation techniques were developed to improve software productivity. HLLs, which are platform independent, follow the rules of human language with a grammar, a syntax, and a semantics. They thus provide flexibility and portability by hiding details of the computer architecture. Assembly language is today used only in limited scenarios, primarily to optimize the critical parts of a program when there is an absolute need for speed and code compactness, or both. However, with the growing complexity of both modern system architectures and software applications, using HLLs and compilers clearly generates better overall results. No one today would even think of programming a complex software application solely by using an assembly language.

In the hardware domain, specification languages and design methodologies have evolved similarly. For this reason, until the late 1960s, ICs were designed, optimized, and laid out by hand. Simulation at the gate level appeared in the early 1970s, and cycle-based simulation became available by 1979. Techniques introduced during the 1980s included place-and-route, schematic circuit capture, formal verification, and static timing analysis. Hardware description languages (HDLs), such as Verilog (1986) and VHDL (1987), have enabled wide adoption of simulation tools. These HDLs have also served as inputs to logic synthesis tools leading to the definition of their synthesizable subsets. During the 1990s, the first generation of commercial high-level synthesis (HLS) tools was available commercially. Around the same time, research interest on hardware-software code sign-including estimation, exploration, partitioning, interfacing, communication, synthesis, and cosimulation-gained momentum. The concept of IP core and platform-based design started to emerge. In the 2000s, there has been a shift to an electronic system-level (ESL) paradigm that facilitates exploration, synthesis, and verification of complex SoCs. This includes the introduction of languages with system-level abstractions, such as SystemC, SpecC, or SystemVerilog, and the introduction of transaction-level modeling (TLM). The ESL paradigm shift caused by the rise of system complexities, a multitude of components in a product (hundreds of processors in a car, for instance), a multitude of versions of a chip (for better product differentiation), and an interdependency of component suppliers forced the market to focus on hardware and software productivity, dependability, interoperability, and reusability. In this context, processor customization and HLS have become necessary paths to efficient ESL design. The new HLS flows, in addition to reducing the time for creating the hardware, also help reduce the time to verify it as well as facilitate other flows such as power analysis

Raising the hardware design's abstraction level is essential to evaluating system-level exploration for architectural decisions such as hardware and software design, synthesis and verification, memory organization, and power management. HLS also enables reuse of the same high-level specification, targeted to accommodate a wide range of design constraints and ASIC or FPGA technologies.

Typically, a designer begins the specification of an application that is to be implemented as a custom processor, dedicated coprocessor or any other custom hardware unit such as interrupt controller, bridge, arbiter, interface unit, or a special function unit with a high-level description capture of the desired functionality, using an HLL. This first step thus involves writing a functional specification (an untimed description) in which a function consumes all its input data simultaneously, performs all computations without any delay, and provides all its output data simultaneously. At this abstraction level, variables (structure and array) and data types (typically floating point and integer) are related neither to the hardware design domain (bits, bit vectors) nor to the embedded software. Realistic hardware implementation thus requires conversion of floating-point and integer data types into bit-accurate data types of specific length (not a standard byte or word size, as in software) with acceptable computation accuracy, while generating an optimized hardware architecture starting from this bit-accurate specification.

HLS tools transform an untimed (or partially timed) high-level specification into a fully timed implementation. They automatically or semi automatically generate a custom architecture to efficiently implement the specification. In addition to the memory banks and the communication interfaces, the generated architecture is described at the RTL and contains a data path (registers, multiplexers, functional units, and buses) and a controller, as required by the given specification and the design constraints.

## VLSI Design

Very-large-scale integration (VLSI) is the process of creating an integrated circuit (IC) by combining thousands of transistors into a single chip. VLSI began in the 1970s when complex semiconductor and communication technologies were being developed. The microprocessor is a VLSI device.

Before the introduction of VLSI technology, most ICs had a limited set of functions they could perform. An electronic circuit might consist of a CPU, ROM, RAM and other glue logic. VLSI lets IC designers add all of these into one chip.

The electronics industry has achieved a phenomenal growth over the last few decades, mainly due to the rapid advances in large scale integration technologies and system design applications. With the advent of very large scale integration (VLSI) designs, the number of applications of integrated circuits (ICs) in high-performance computing, controls, telecommunications, image and video processing, and consumer electronics has been rising at a very fast pace.

The current cutting-edge technologies such as high resolution and low bit-rate video and cellular communications provide the end-users a marvelous amount of applications, processing power and portability. This trend is expected to grow rapidly, with very important implications on VLSI design and systems design.

## VLSI Design Flow

The VLSI IC circuits design flow is shown in the figure below. The various levels of design are numbered and the blocks show processes in the design flow.

Specifications comes first, they describe abstractly, the functionality, interface, and the architecture of the digital IC circuit to be designed.

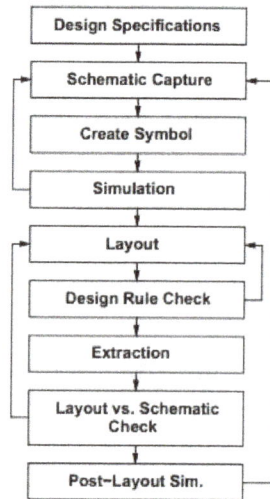

Behavioral description is then created to analyze the design in terms of functionality, performance, compliance to given standards, and other specifications.

RTL description is done using HDLs. This RTL description is simulated to test functionality. From here onwards we need the help of EDA tools.

RTL description is then converted to a gate-level net list using logic synthesis tools. A gate level net list is a description of the circuit in terms of gates and connections between them, which are made in such a way that they meet the timing, power and area specifications.

Finally, a physical layout is made, which will be verified and then sent to fabrication.

## Y Chart

The Gajski-Kuhn Y-chart is a model, which captures the considerations in designing semiconductor devices.

The three domains of the Gajski-Kuhn Y-chart are on radial axes. Each of the domains can be divided into levels of abstraction, using concentric rings.

At the top level (outer ring), we consider the architecture of the chip; at the lower levels (inner rings), we successively refine the design into finer detailed implementation –

Creating a structural description from a behavioral one is achieved through the processes of high-level synthesis or logical synthesis.

Creating a physical description from a structural one is achieved through layout synthesis.

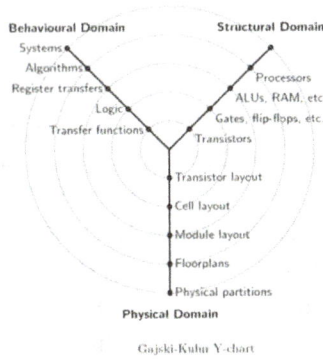

Gajski-Kuhn Y-chart

## Design Hierarchy-structural

The design hierarchy involves the principle of "Divide and Conquer." It is nothing but dividing the task into smaller tasks until it reaches to its simplest level. This process is most suitable because the last evolution of design has become so simple that its manufacturing becomes easier.

We can design the given task into the design flow process's domain (Behavioral, Structural, and Geometrical). To understand this, let's take an example of designing a 16-bit adder, as shown in the figure below.

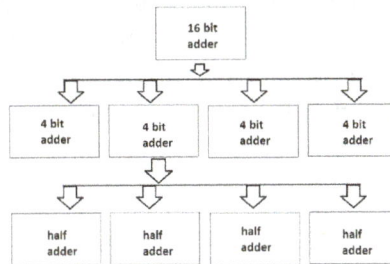

Here, the whole chip of 16 bit adder is divided into four modules of 4-bit adders. Further, dividing the 4-bit adder into 1-bit adder or half adder. 1 bit addition is the simplest designing process and its internal circuit is also easy to fabricate on the chip. Now, connecting all the last four adders, we can design a 4-bit adder and moving on, we can design a 16-bit adder.

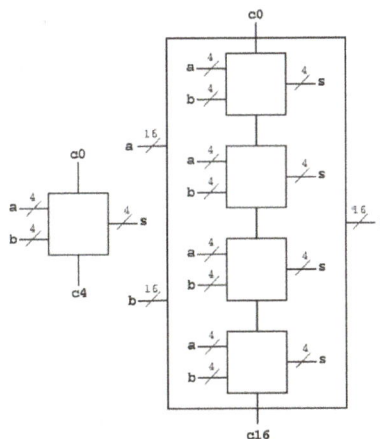

# Low Power VLSI Design

During the desktop PC design era VLSI design efforts have focused primarily on optimizing speed to realize computationally intensive real-time functions such as video compression, gaming, graphics etc. As a result, we have semiconductor ICs that successfully integrated various complex signal processing modules and graphical processing units to meet our computation and entertainment demands. While these solutions have addressed the real-time problem, they have not addressed the increasing demand for portable operation, where mobile phone need to pack all this without consuming much power. The strict limitation on power dissipation in portable electronics applications such as smart phones and tablet computers must be met by the VLSI chip designer while still meeting the computational requirements. While wireless devices are rapidly making their way to the consumer electronics market, a key design constraint for portable operation namely the total power consumption of the device must be addressed. Reducing the total power consumption in such systems is important since it is desirable to maximize the run time with minimum requirements on size, battery life and weight allocated to batteries. So the most important factor to consider while designing System on Chip (SoC) for portable devices is 'low power design'.

Scaling of technology node increases power-density more than expected. CMOS technology beyond 65nm node represents a real challenge for any sort of voltage and frequency scaling Starting from 120nm node, each new process has inherently higher dynamic and leakage current density with minimal improvement in speed. Between 90nm to 65nm the dynamic power dissipation is almost same whereas there is ~5% higher leakage/mm2.Low cost always continues to drive higher levels of integration, whereas low cost technological breakthroughs to keep power under control are getting very scarce.

Modern System-on-Chip demand for more power. In both logic and memory, Static power is growing really fast and Dynamic power kind of grows. Overall power is dramatically increasing. If the semiconductor integration continue to follow Moore's Law, the power density inside the chips will reach far higher than the rocket nozzle.

Power dissipation is the main constrain when it comes to Portability. The mobile device consumer demands more features and extended battery life at a lower cost. About 70% of users demand longer talk and stand-by time as primary mobile phone feature. Top 3G requirement for operators is power efficiency. Customers want smaller & sleeker mobile devices. This requires high levels of Silicon integration in advanced processes, but advanced processes have inherently higher leakage current. So there is a need to bother more on reducing leakage curret to reduce power consumption.

## Importance of Power in SOC

Power Management matter in System on Chip due to following concerns

- a) Packaging and Cooling costs.
- b) Digital noise immunity.

c) Battery life (in portable systems).

d) Environmental concerns.

## Sources of Power Dissipation

The power dissipation in circuit can be classified into three categories as described below.

Dynamic power consumption: Due to logic transitions causing logic gates to charge/discharge load capacitance.

Short-circuit current: In a CMOS logic P-branch and N-branch are momentarily shorted as logic gate changes state resulting in short circuit power dissipation.

Leakage current: This is the power dissipation that occurs when the system is in standby mode or not powered. There are many sources of leakage current in MOSFET. Diode leakages around transistors and n-wells, Sub threshold Leakage, Gate Leakage, Tunnel Currents etc. Increasing 20 times for each new fabrication technology. Went from insignificant to a dominating factor.

## Low-power Design Techniques

An integrated low power methodology requires optimization at all design abstraction layers as mentioned below.

- System: Partitioning, Power down

- Algorithm: Complexity, Concurrency, Regularity

- Architecture: Parallelism, Pipelining, Redundancy, Data Encoding

- Circuit Logic: Logic Styles, Energy Recovery, Transistor Sizing

- Technology: Threshold Reduction, Multithreshold Devices.

Dynamic power varies as VDD2. So reducing the supply voltage reduces power dissipation. Also selective frequency reduction technique can be used to reduce dynamic power. Multi threshold voltage can be used to reduce leakage power at system level. Transistor resizing can be used to speed-up circuit and reduce power. Sleep transistors can be used effectively to reduce standby power. Parallelism and pipelining in system architecture can reduce power significantly. Clock disabling, power-down of selected logic blocks, adiabatic computing, software redesign to lower power dissipation are the other techniques commonly used for low power design.

## VLSI Circuit Design for Low Power

The growing market of portable (e.g., cellular phones, gaming consoles, etc.), battery-powered electronic systems demands microelectronic circuits design with ultra-low power dissipation. As the integration, size, and complexity of the chips continue to increase, the difficulty in providing adequate cooling might either add significant cost or limit the functionality of the computing systems which make use of those integrated circuits. As the technology node scales down to 65 nm there is not much increase in dynamic power dissipation. However the static or leakage power is same as or exceeds the dynamic power beyond 65 nm technology node.

Hence the techniques to reduce power dissipation is not limited to dynamic power. In this topic we discuss circuit and logic design approaches to minimize Dynamic, Leakage and Short Circuit power dissipation. Power optimization in a processor can be achieved at various abstract levels. System/Algorithm/Architecture has a large potential for power saving even these techniques tend to saturate as we integrate more functionality on an IC. So optimization at Circuit and Technology level is also very important for miniaturization of ICs.

Total Power dissipated in a CMOS circuit is sum total of dynamic power, short circuit power and static or leakage power. Design for low-power implies the ability to reduce all three components of power consumption in CMOS circuits during the development of a low power electronic product. We summarize the most widely used circuit techniques to reduce each of these components of power in a standard CMOS design.

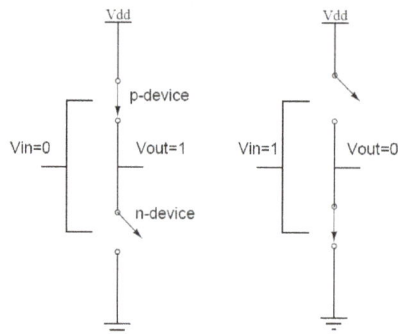

Figure: Components of Power in CMOS circuit

Ptotal = CLVDD2 + tscVDDIpeak + VDDIleakage

## Dynamic Power Suppression

Dynamic/Switching power is due to charging and discharging of load capacitors driven by the circuit. Supply voltage scaling has been the most adopted approach to power optimization, since it normally yields considerable power savings due to the quadratic dependence of switching/dynamic power PSwitching on supply voltage VDD. However lowering the supply voltage affects circuit speed which is the major short-coming of this approach. So both design and technological solutions must be applied to compensate the decrease in circuit performance introduced by reduced voltage. Some of the techniques often used to reduce dynamic power are described below.

## Adiabatic Circuits

In adiabatic circuits instead of dissipating the power is reused. By externally controlling the length and shape of signal transitions energy spent to flip a bit can be reduced to very small values. Since diodes are thermodynamically irreversible they are not used in the design of Adiabatic Logic. MOSFETs should not be turned ON when there is significant potential difference between source and drain. And should not be turnoff when there is a significant current flowing through the device.

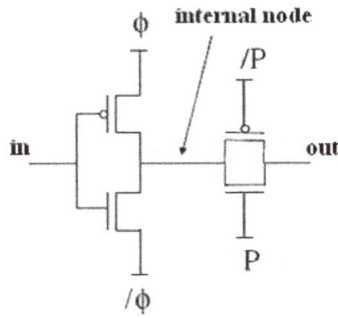

Figure: Charge Recovery Logic

In the adiabatic circuit shown above initially, f and/f at Vdd/2, P at Gnd, and/P at Vdd. On valid input, the pass gate is turned on by gradually swinging P and/P. Rails f and/f "split", gradually swinging to Vdd and Gnd. As soon as output is sampled, pass gate is turned off. Internal node is restored by gradually swinging f and/f back to Vdd/2.Once the device is on energy transfer takes place in a controlled manner so that there is no potential drop across the device.

## Logic Design for Low Power

Choices between static versus dynamic topologies, conventional CMOS versus pass-transistor logic styles and synchronous versus asynchronous timing styles have to be made during the design of a circuit. In static CMOS circuits, the component of power due to short circuit current is about the 10% of the total power consumption. However, in dynamic circuits we don't come across this problem, since there is no any direct dc path from supply voltage to ground. Only in domino-logic circuits there is such a path, in order to reduce sharing, hence there is a small amount of short-circuit power dissipation.

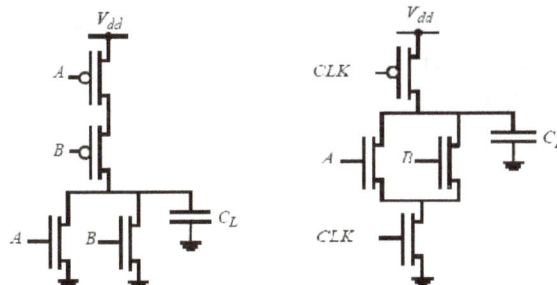

Figure: (a) Static NOR and    Figure: (b) Dynamic NOR circuits

Similarly we can use pass transistor logic to exploit reduced swing to lower power (e.g., reduced bit-line swing in memory).

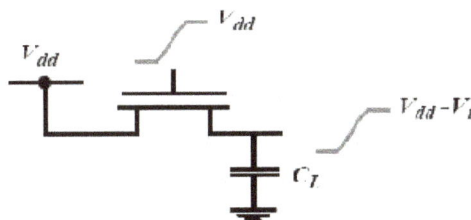

Figure: Pass Transistor Logic

$$P = CL* Vdd* (Vdd-Vt)$$

## Reducing Glitches

Glitches occur in a logic chain when two parallel driving common gate arrive at different times. The output momentarily switches to incorrect value before settling to correct result. Consider circuit shown below. Let us assume that in the absence of buffer path A is high speed and Path B is slow. Initially if A=0 and B=1 then Z=0. Next if B is to switch to 0 and A to 1 since B is slow the data 0 arriving at B will be slow and hence Z switches towards 1 momentarily before switching back to 0 resulting in power dissipation.

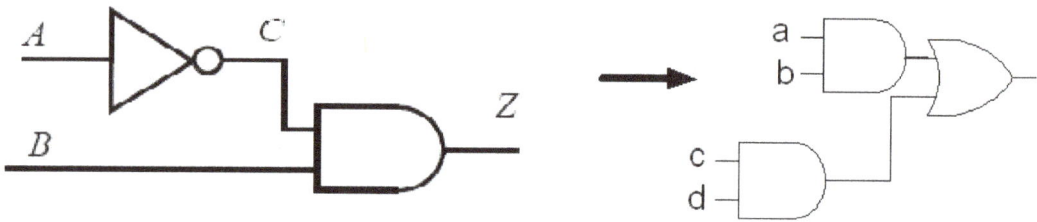

Figure: Glitch Free AND Gate

As shown in figure above buffers are generally used to delay path A to overcome glitches.

## Logic Level Power Optimization

During logic optimization for low power, technology parameters such as supply voltage are fixed, and the degrees of freedom are in selecting the functionality and sizing the gates. Path equalization with buffer insertion is one of the techniques which ensure that signal propagation from inputs to outputs of a logic network follows paths of similar length to overcome glitches. When paths are equalized, most gates have aligned transitions at their inputs, thereby minimizing spurious switching activity/glitches (which are created by misaligned input transitions).

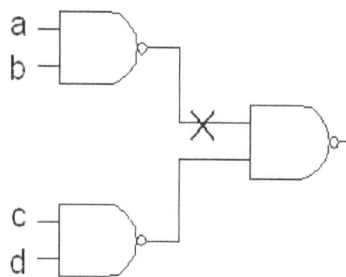

Figure: Logic Remapping for Low Power

Other logic-level power minimization techniques include local transformations as shown in figure above. A re-mapping transformation is shown, where a high-activity node (marked with x) is removed and replaced by new mapping onto an and or gate.

## Standby Mode Leakage Suppression

Static/Leakage power, originates from substrate currents and sub threshold leakages. For technologies 1 μm and above, PSwitching was predominant. However for deep-submicron processes below 180nm, PLeakage becomes dominant factor. Leakage power is a major concern in recent technologies, as it impacts battery lifetime. CMOS technology has been extremely power-efficient when transistors are not switching or in stand-by mode, and system designers expect low leakage from CMOS chips. To meet leakage power constraints, multiple-threshold and variable threshold circuit techniques are often used. In multiple-threshold CMOS, the process provides two different threshold transitors. Low-threshold are employed on speed-critical sub-circuits and ther are fast and leaky. High-threshold transistors are slower but exhibit low sub-threshold leakage, and they are employed in noncritical/slow paths of the chip. As more transistors become timing-critical multiple-threshold techniques tend to lose effectiveness.

## Variable Body Biasing

Variable-threshold circuits dynamically control the threshold voltage of transistors through substrate biasing and hence overcome shortcoming associated with multi-threshold design. When a variable-threshold circuit is in standby, the substrate of NMOS transistors is negatively biased, and their threshold increases because of the body-bias effect. Similarly the substrate of PMOS transistors is biased by positive body bias to increase their Vt in stand-by. Variable-threshold circuits can, in principle, solve the quiescent/static leakage problem, but they require control circuits that modulate substrate voltage in stand-by. Fast and accurate body-bias control with control circuit is quite challenging, and requires carefully designed closed-loop control. When the circuit is in standby mode the bulk/body of both PMOS and NMOS are biased by third supply voltage to increase the Vt of the MOSFET as shown in the Figure. However during normal operation they are switched back to reduce the Vt.

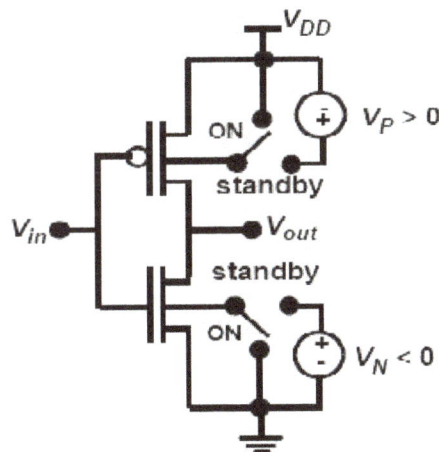

Figure: Variable Body Biasing

## Sleep Transistors

Sleep Transistors are High Vt transistors connected in series with low Vt logic as shown below. When the main circuit consisting of Low Vt devices are ON the sleep transistors are also ON resulting in normal operation of the circuit. When the circuit is in Standby mode even High Vt transistors are OFF. Since High Vt devices appear in series with Low Vt circuit the leakage current is determined by High Vt devices and is very low. So the net static power dissipation is reduced.

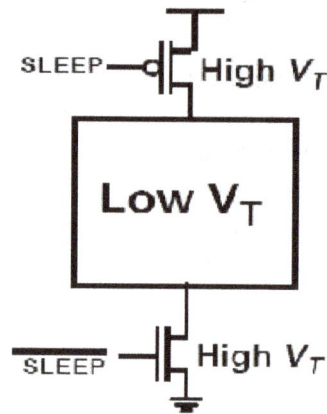

Figure: Circuit Design with Sleep Transistors

## Dynamic Threshold MOS

In dynamic threshold CMOS (DTMOS), the threshold voltage is altered dynamically to suit the operating state of the circuit. A high threshold voltage in the standby mode gives low leakage current, while a low threshold voltage allows for higher current drives in the active mode of operation. Dynamic threshold CMOS can be achieved by tying the gate and body together. The supply voltage of DTMOS is limited by the diode built-in potential in bulk silicon technology. The pn diode between source and body should be reverse biased. Hence, this technique is only suitable for ultralow voltage (0.6V and below) circuits in bulk CMOS.

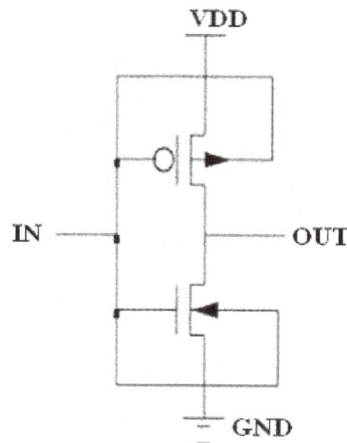

Figure: DTMOS Circuit

## Short Circuit Power Suppression

Short-circuit power, is caused by the short circuit currents that arise when pairs of PMOS/NMOS transistors are conducting simultaneously. In static CMOS circuits, short-circuit path exists for direct current flow from VDD to ground, when

$$VTn < Vin < VDD - |VTp|$$

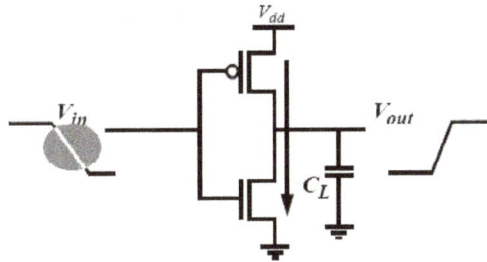

Figure: Short Circuit Power in CMOS Circuits

One way to reduce short circuit power is to keep the input and output rise/fall times the same. If Vdd < Vtn + |Vtp| then short-circuit power can be eliminated. If the load capacitance is very large, the output fall time is larger than the input rise time. The drain-source voltage of the PMOS transistor is 0. Hence the short-circuit power will be 0. If the load capacitance is very small, the output fall time is smaller than the input rise time. The drain-source voltage of the PMOS transistor is close to VDD during most of the transition period. Hence the short-circuit power will be very large.

## Interconnect Delay

Interconnect delay is also known as net delay or wire delay or extrinsic delay or flight time. Interconnect delay is the difference between the time a signal is first applied to the net and the time it reaches other devices connected to that net.

It is due to the finite resistance and capacitance of the net. It is also known as wire delay.

$$\text{Wire delay} = \text{function of (Rnet, Cnet+Cpin)}$$

This is output pin of the cell to the input pin of the next cell.

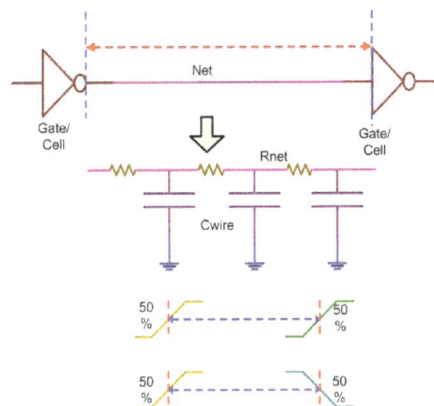

Net delay is calculated using Rs and Cs.

There are several factors which affect net parasitic:

- Net Length
- Net cross-sectional area

- Resistively of material used for metal layers (Aluminum vs. copper)

- Number of vias traversed by the net

- Proximity to other nets (crosstalk)

Post-layout design is annotated with RCs extracted from layout for better accuracy. Annotated RCs override information from WLM.

Interconnect introduces capacitive, resistive and inductive parasites. All three have multiple effects on the circuit behavior.

1. Interconnect parasites cause an increase in propagation delay (i.e. it slows down working speed)

2. Interconnect parasites increase energy dissipation and affect the power distribution.

3. Interconnect parasites introduce extra noise sources, which affect reliability of the circuit. (Signal Integrity effects)

Dominant parameters determine the circuit behavior at a given circuit node. Non-dominant parameters can be neglected for interconnect analysis.

- Inductive effect can be ignored if the resistance of the wire is substantial enough-this is the case for long aluminum wires with a small cross section or if the rise and fall times of the applied signals are slow.

- When the wires are short, the cross section of the wire is large or the interconnect material used has a low resistivity, a capacitive only model can be used.

- When the separation between neighboring wires is large or when the wires only run together for short distance, inter-wire capacitance can be ignored, and all the parasitic capacitance can be modeled as capacitance to ground.

## Lumped Capacitor Model

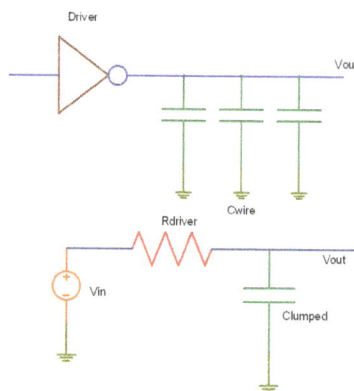

As long as the resistive component of the wire is small, and switching frequencies are in the low to medium range, it is meaningful to consider only the capacitive component of the wire, and to lump the distributed capacitance into a single capacitance.

The only impact on performance is introduced by the loading effect of the capacitor on the driving gate.

## Lumped RC Model

If wire length is more than a few millimeters, the lumped capacitance model is inadequate and a resistive capacitive model has to be adopted.

In lumped RC model the total resistance of each wire segment is lumped into one single R, combines the global capacitive into single capacitor C.

Analysis of network with larger number of R and C becomes complex as network contains many time constants (zeroes and poles). Elmore delay model overcome such problem.

## Elmore Delay Model

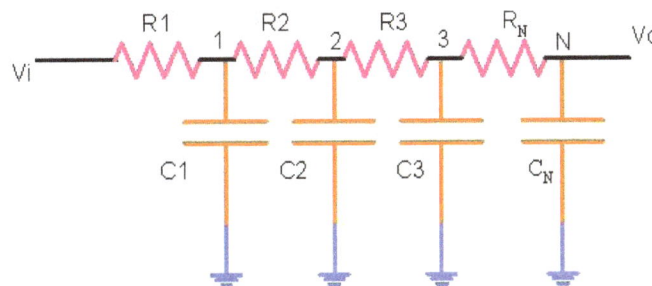

Properties of the network:

- Has single input node
- All the capacitors are between a node and ground.
- Network does not contain any resistive loops.

"Path resistance" is the resistance from source node to any other node.

"Shared path resistance" is the resistance shared among the paths from the source node to any other two nodes.

Hence,

Delay at node 1: Tow $d_1 = R_1C_1$

Delay at node 2: Tow $d_2 = (R_1+R_2)C_2$

Delay at node 3: Tow $d_3 = (R_1+R_2+R_3)C_3$

In general:

$$T_{di} = R_1C_1 + (R_1+R_2)C_2 + \ldots\ldots + (R_1+R_2+R_3+\ldots\ldots+R_i)C_i$$

If

$$R_1 = R_2 = R_3 = .... = R$$

$$C_1 = C_2 = C_3 = ..... C \text{ then}$$

$$T_{di} = 1RC + 2RC + ........ + nRC$$

Thus Elmore delay is equivalent to the first order time constant of the network.

Assuming an interconnect wire of length L is partitioned into N identical segments. Each segment has length L/N.

Then,

$$T_d = L/N.R.L/N.C + 2 (L/n.r + L/N.C) + ......$$

$$= (L/N)2(RC + 2RC + ....... + NRC)$$

$$= (L/N)2. N(N+1)$$

or $T_d = RC.L2/2$

=> The delay of a wire is a quadratic function of its length

=> doubling the length of the wire quadruples its delay

## Advantages

- It is simple
- It is always situated between minimum and maximum bounds

## Disadvantages

- It is pessimistic and inaccurate for long interconnect wires

## Distributed RC Model

Lumped RC model is always pessimistic and distributed RC model provides better accuracy over lumped RC model.

But distributed RC model is complex and no closed form solution exists. Hence distributed RC line model is not suitable for Computer Aided Design Tools.

The behavior of the distributed RC line can be approximated by a lumped RC ladder network such as Elmore Delay model hence these are extensively used in EDA tools.

## Transmission Line Model

When frequency of operation increases to a larger extent, rise (or fall) time of the signal becomes comparable to time of flight of the net, then inductive effects starts dominating over RC values.

This inductive effect is modeled by Transmission Line models. The model assumes that the signal is a "wave" and it propagates over the medium "net".

There are two types of transmission models:

- Lossless transmission line model: This is good for Printed Circuit Board level design.

- Lossy transmission line model: This model is used for IC interconnect model.

Transmission line effects should be considered when the rise or fall time of the input signal is smaller than the time of flight of the transmission line or resistance of the wire is less than characteristics impedance.

## Wire Load Models

Extraction data from already routed designs are used to build a lookup table known as the wire load model (WLM). WLM is based on the statistical estimates of R and C based on "Net Fan-out".

For fan outs greater than those specified in a wire load table, a "slope factor" is specified for linear extrapolation.

wire_load ("5KGATES") {

resistance : 0.000271 --------------> R per unit length

capacitance : 0.00017 --------------> C per unit length

slope : 29.4005 ----------------------> Used for linear extrapolation

fanout_length (1, 18.38) ----------> (fanout = 1, length = 18.38)

fanout_length (2, 47.78)

fanout_length (3, 77.18)

fanout_length (4, 106.58)

fanout_length (5, 135.98)

}

Eg:

Fanout = 7

Net length = 135.98 + 2 x 29.4005 (slope) = 194.78 ---------->

length of net with fanout of 7

Resistance = 194.78 x 0.000271 = 0.05279 units

Capacitance = 194.78 x 0.00017 = 0.03311 units

Wire load models for synthesis

Wire load modeling allows us to estimate the effect of wire length and fanout on the resistance, capacitance, and area of nets. Synthesizer uses these physical values to calculate wire delays and circuit speeds. Semiconductor vendors develop wire load models, based on statistical information specific to the vendors' process. The models include coefficients for area, capacitance, and resistance per unit length, and a fanout-to-length table for estimating net lengths (the number of fanouts determines a nominal length).

Selection of wire load models in the initial stage (before physical design) depends on the fallowing factors:

1. User specification

2. Automatic selection based on design area

3. Default specification in the technology library

Once the final routing step is over in the physical design stage, wire load models are generated based on the actual routing in the design and synthesis is redone using those wire load models.

In hierarchical designs, we have to determine which wire load model to use for nets that cross hierarchical boundaries. There are three modes for determining which wire load model to use for nets that cross hierarchical boundaries:

## Top

Applying same wire load models to all nets as if the design has no hierarchy and uses the wire load model specified for the top level of the design hierarchy for all nets in a design and its sub designs.

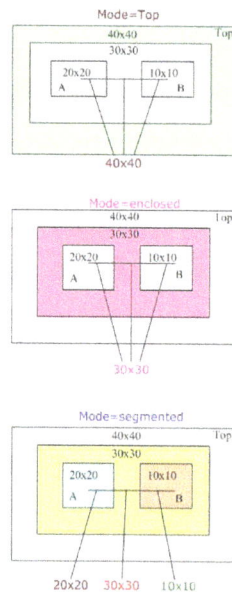

## Enclosed

The wire load model of the smallest design that fully encloses the net is applied. If the design enclosing the net has no wire load model, then traverses the design hierarchy upward until we finds a wire load model. Enclosed mode is more accurate than top mode when cells in the same design are placed in a contiguous region during layout.

Use enclosed mode if the design has similar logical and physical hierarchies.

## Segmented

Wire load model for each segment of a net is determined by the design encompassing the segment. Nets crossing hierarchical boundaries are divided into segments. For each net segment, the wire load model of the design containing the segment is used. If the design contains a segment that has no wire load model, then traverse the design hierarchy upward until it finds a wire load model.

## Interconnect Delay vs. Deep Sub Micron Issues

Performances of deep sub-micron ICs are limited by increasing interconnect loading affect. Long global clock networks account for the larger part of the power consumption in chips. Traditional CAD design methodologies are largely affected by the interconnect scaling. Capacitance and resistance of interconnects have increased due to the smaller wire cross sections, smaller wire pitch and longer length. This has resulted in increased RC delay. As technology is advancing scaling of interconnect is also increasing. In such scenario increased RC delay is becoming major bottleneck in improving performance of advanced ICs.

Here the gate delay and the interconnect delay are shown as functions of various technology nodes ranging from 180nm to 60nm. The interconnect delays shown assumes a line where repeaters are connected optimally and includes the delay due to the repeaters. From the graph it can be observed that with the shrinking of technology gate delay reduces but interconnect delay increases.

## Limits of Cu/low-k Interconnects

At submicron level of 250 nm copper with low-k dielectric was introduced to decrease effects of increasing interconnect delay. But below 130 nm technology node interconnect delays are increasing further despite of introducing low-k dielectric. As the scaling increases new physical and technological effects like resistivity and barrier thickness start dominating and interconnect delay increases. Introduction of repeaters to shorten the interconnect length increases total area. The vias connecting repeaters to global layers can cause blockage in lower metal layers. Thus as the technology improves material limitations will dominate factor in the interconnect delay. Increasing metal layer width will cause increase in metallization layer. This can't be a solution for the problem as it increases complexity, reliability and cost.

Cu low-k dielectric films are deposited by a special process known as Damascene process. Adhesion property of Cu with dielectric materials is very poor. Under electric bias they easily drift and cause short between metal layers. To avoid this problem a barrier layer is deposited between dielectric and Cu trench. Even though it decreases effective cross section of interconnects compared to drawn dimensions, it improves reliability. The barrier thickness becomes significant in deep submicron level and effective resistance of the interconnect rises further. In addition to this increasing electron scattering and self-heating caused by the electron flow in interconnects due to comparable increase in internal chip temperature also contribute to increase interconnect resistance.

# Design for Manufacturability

Aggressive ground rule changes continue to increase the complexity of semiconductor technology. The requirements for designs, processes, equipment, and facilities all grow in sophistication from generation to generation. These trends have made it increasingly difficult to produce a technology in the development laboratory and transfer it to volume manufacturing in a timely and cost effective manner. The traditional laboratory role of design and process development has expanded to include a parallel responsibility for manufacturability. For many companies, design for manufacture (DFM) has become a critical strategy for survival in an increasingly competitive global marketplace. DFM is a systems approach to improving the competitiveness of a manufacturing enterprise by developing products that are easier, faster, and less expensive to make, while maintaining required standards of functionality, quality, and marketability. Design for manufacturability (DFM) and early manufacturing involvement (EMI) concepts are now major components of the development effort designed to maintain and enhance the rate of technology advancement and significantly improve the development-to-manufacturing transition. Design-for-manufacturability philosophy and practices are used in many companies because it is recognized that 70% to 90% of overall product cost is determined before a design is ever released into manufacturing. The semiconductor industry continues to grow in both complexity and competitiveness.

## Problem Statement

The layout development is most critical in integrated circuits (IC's) design because of cost, since it involves expensive tools and a large amount of human intervention, and also because of the consequences for production cost. As the device size is shrinking, the landscape of technology developments has become very different from the past. The problems, which were supposed to be secondary can cause of yield drop out in submicron technologies. The variability becomes a critical issue not only for performance, but also for yield dropout.

Yield dropout due to given below defects.

Random Defects: Due to form of impurities in the silicon itself, or the introduction of a dust particle that land on the wafer during processing. These defects can cause a metal open or shorts. As feature sizes continue to shrink, random defects have not decreased accordingly making advanced IC's even more susceptible to this type of defect.

Systematic Defects: Again systematic defects are more prominent contributor in yield loss in deep submicron process technologies. Systematic defects are related to process technology due to limitation of lithography process which increased the variation in desired and printed patterns. Another aspects of process related problem is planarity issues make layer density requirements necessary because areas with a low density of a particular layer can cause upper layers to sag, resulting in discontinuous planarity across the chip.

Parametric Defects: In deep submicron technology parametric defects is most critical for us. Parametric defects come into the picture due to improper modeling of interconnects parasitic. As a result manufactured device does not match the expected result from design simulation and does not meet the design specification.

Design for manufacturability (DFM) is process to overcome these defects of yield drop out. The DFM will not be done without collaborations between various technology parties, such as process, design, mask, EDA, and so on. The DFM will give us a big challenge and opportunity in nanometer era.

Design for Manufacturability is the proactive process which ensures the quality, reliability, cost effective and time to market.

DFM consist a set of different methodologies trying to enforce some soft (recommended/Mandatory) design rules regarding the shapes and polygons of the physical layout which improve the yield.

Given a fixed amount of available space in a given layout area, there are potentially multiple yield enhancing changes that can be made.

There are some DFM guidelines which we can take into account during layout.

- Antenna effect guidelines

  Configuration: Identify poly gates connected to large areas of metals.

  Action: Reduce the ratio surface of metal/surface poly gate or use free wheel diode.

  Reasons for this action: During process, the wafers are submitted to plasma environments. Because of metallization, electrons are collected. Because of mirror effects charges accumulate at the gate oxide. An electric field is created and can cause oxide breakdown.

- Minimum Area of physical layers

  Configuration: Identify very small rectangle of a given layer (typically shape at the min area size like diffusion)

  Action: Try not to draw shapes at min area when free space is available around.

  Reasons for this action: Process window can allow shapes to be at min area, but if these shapes are numerous, risk is higher that some of them are not resolved: for instance, in case of implant, that would lead to missing implant. Typical case: Pwell and Nwell Straps (Ties).

- Density Gradient

  Configuration: Identify high density areas next to low density areas.

  Action: Try to balance shapes to reach homogeneous density and add dummy patterns.

  Reasons for this action: Density of some layers (specifically those treated by Chemical & Mechanical Polishing (CMP) like diffusion, poly and metals) has a big impact on the manufacturing of the given pattern. Impact is both on CMP processing and photo processing. Too high gradient can lead to shorts or opens.

- Contact Enclosure by Diffusion/poly silicon

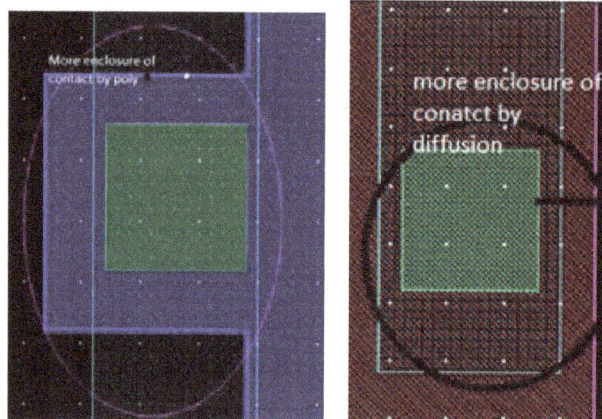

Configuration: Identify min enclosure of contacts by diffusion.

Action: Try to extend the enclosure of contact by diffusion area when possible.

Reasons for this action: Overlay could make that one contact falls on the border of the diffusion area, thus generating a junction leakage.

Caution: Proportions of the dimensions of this transistor have not been kept, for a better visibility of the example. Take care of resulting increase of drain capacitance.

- Metal extension of Via/contact at Line Ends

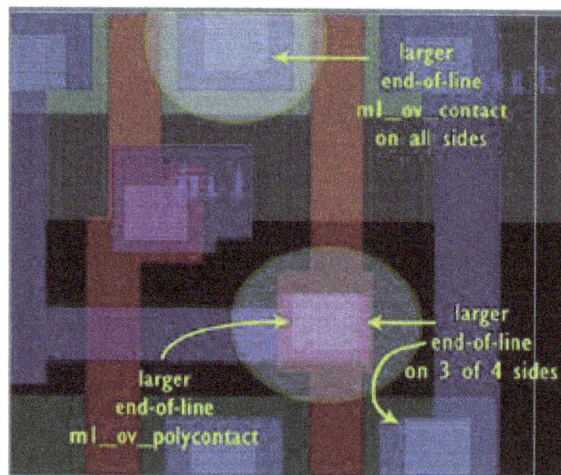

Configuration: Identify Via transitions at line ends.

Action: Try to extend the overlap of metals at line ends.

Reasons for this action: Process window issues might lead to resistive or even open vias. Extending metals overlap ensures a better transition.

- Gate Extension on Diffusion (End Cap)

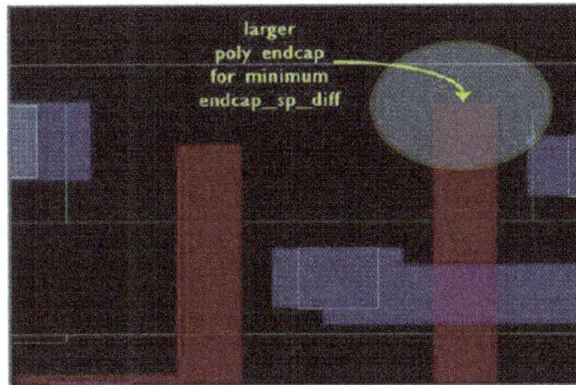

- Configuration: Identify poly gate of transistor at min from diffusion.

- Action: Try to extend the poly end cap wherever it is not too close to other structures

- Reasons for this action: Silicon implementation of this configuration could lead to leakage current between drain and source of the given transistor. Do not draw at min if space is available.

- Contact/Via Redundancy

  Configuration: Identify single contact specifically for critical transistor in repetitive cell.

  Action: Try to double contact and extend poly and metal1 without impacting too much on poly & metal1 critical area.

  Reasons for this action: Single contacts can be more sensitive to defect and resistivity spread especially in case of L transition. Silicide discontinuity risk is also present. Inserting a new contact reduces probability to have a too resistive (or even open) transition, and reduces electro-migration effect.

- Metal to Metal spacing

  Configuration: Identify wires at min spacing with free space around

  Action: Do not keep min spacing. Try to spread these wires and if not possible decrease width/space ratio.

  Reasons for this action: Long wires at min spacing increase probability to have shorts due to particles

- Via/contact to Via/contact spacing

  Configuration: Identify large via to via spacing.

  Action: Decrease Via to via spacing by adding dummy vias.

  Reasons for this action: Lone Vias or edge Vias tend to be gathering points of Low K outgassing. This results in resist poisoning and poor Vias lithography and etching => Via Opens

  Dummy vias must be added above metal lines or dummies.

Regular layouts

There are some DFM guidelines which we can take into account at SOC level.

1.  Filler cell (consisting regular Diffusion and Poly silicon structures) insertion and shielding

    Issue Addressed: PO/OD non uniformity

    Benefit: Higher parametric yield.

2.  Via optimization

    Issue Addressed: open Via's systematic via opening issue

    Benefit: Higher yield after manufacturing and qualification.

3.  Wire Spreading

    Issue Addressed: wire shorts and opening due to defectivity.

    Benefit: Higher yield, decrease cross talk.

4.  Power/ground-connected fill

    Issue Addressed: Density gradients, Large IR drop, Layout becomes regular

    Benefit: Robustness to IR drop

5.  Litho hotspot detection and repair

    Issue Addressed: Lithography hotspots

    Benefit: Higher yield

6.  Dummy Metal/Via/FEOL

    Issue Addressed: Large density gradients

    Benefit: Higher yield

7.  CMP hotspot detection

    Issue Addressed: CMP hotspots

    Benefit: Higher yield

## Standard Cell

A gate array is a fixed array of cells containing transistors and (in some technologies) resistors. Cells vary in size depending upon the vendor and technology. Each internal cell however is the same as every other internal cell in the array. I/O cells differ from the array cells because of drive and input protection reasons, but again, all I/O cells are identical. There are inefficiencies in gate utilization because of bus crossers, and the interconnect problem increases as the number of interconnecting gates increase. Large logic blocks such as RAM, ROM, and microprocessor cannot be implemented in gate arrays.

Full custom is the classical approach to integrated circuit design. The majority of the circuit is designed by hand. Full custom has the highest transistor density. However, this is a long, error prone process, and it carries the highest development risk.

Standard cell designs have some of the features of both of the other two methods without their drawbacks. Development times comparable to gate arrays and high densities comparable to full custom can be achieved using standard cells. Standard cells have simple (primitive) cells similar to the gate array. In addition, there are larger more complex combinations of primitives (sub macros and macros). These sub macros and macros have custom interconnect, which contributes to their high densities.

## Design Cycle

## CAD Interplay

To begin the design cycle the designer must be intimately familiar with the standard cell library and the specifications of each entity. He then proceeds with his design based on the available hardware and the circuit specification.

The design cycle flow chart is shown in figure. The architecture of the system is designed by the system designer with testability in mind. After the architecture design is complete, a behavioral level simulation is performed to determine whether or not the design will do what it was intended to do. If the results do not reflect the initial concept, the architecture is changed and simulated again. once the simulation agrees with the concept, the VLSI designer breaks the architecture down to the logic level.

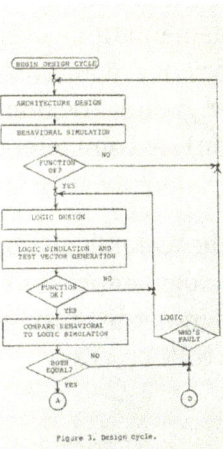

Figure: Design cycle.

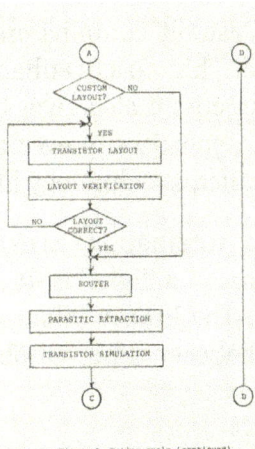

Figure: Design cycle (continued)

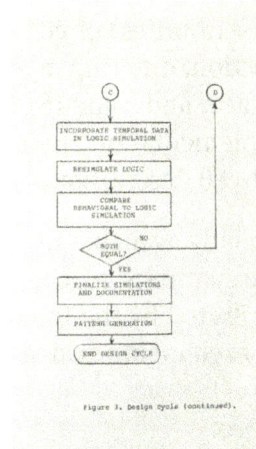

Figure: Design cycle (continued)

A logic simulator is used for the detailed simulation. The delay associated with each device reflects the device delay with nominal interconnection. The delay may be 1.4 times the actual delay with one gate load. This 1.4 factor is due to an average load for each gate of 1.4 loads. Once the logic simulation is complete, it is compared with the behavioral simulation. If they do not correlate, the logic design is checked for errors and changed if any are found. If the problem is in the behavioral simulation, then the hardware is inadequate to support the function. It is obviously assumed that the errors in simulation are weeded out before these decisions are made. If the two simulations do correlate then the physical transistor layout is started.

All of the logical devices used in the logic simulation are available or can be built from the available standards cells in the standard cell library. As the circuit is being laid out, the layout is continually being checked for design rule errors. such errors include signal paths too close to each other, or to active areas of transistors.

The layout is electrically checked for shorted power busses, and open connections. Finally the layout is checked against the logic the verification process characteristic extraction is interconnect resistances simulation model is complete, performed. All After temporal of the and capacitances are automatically extracted from the layout and used for delay calculations. These values are then put into the delay statements of the logic simulation model. The logic simulation is run again and checked against the behavioral simulation. At this point if the simulations do not agree they are checked to see if it was the behavioral or logic simulation at fault. If it was the logic simulation the logic design is approached again. It was the behavioral simulation, the hardware is not adequate and the architecture must be modified.

If the simulations still agree the simulation files are finalized along with the final specifications. The layout then goes to pattern generation (PG).

## Application of Standard Cell

Strictly speaking, a 2-input NAND or NOR function is sufficient to form any arbitrary Boolean function set. But in modern ASIC design, standard-cell methodology is practiced with a sizable library (or libraries) of cells. The library usually contains multiple implementations of the same logic function, differing in area and speed. This variety enhances the efficiency of automated synthesis, place, and route (SPR) tools. Indirectly, it also gives the designer greater freedom to perform implementation trade-offs (area vs. speed vs. power consumption). A complete group of standard-cell descriptions is commonly called a technology library.

Commercially available Electronic Design Automation (EDA) tools use the technology libraries to automate synthesis, placement, and routing of a digital ASIC. The technology library is developed and distributed by the foundry operator. The library (along with a design net list format) is the basis for exchanging design information between different phases of the SPR process.

## Synthesis

Using the technology library's cell logical view, the Logic Synthesis tool performs the process of mathematically transforming the ASIC's register-transfer level (RTL) description into a technology-dependent net list. This process is analogous to a software compiler converting a high-level C-program listing into a processor-dependent assembly-language listing.

The net list is the standard-cell representation of the ASIC design, at the logical view level. It consists of instances of the standard-cell library gates, and port connectivity between gates. Proper synthesis techniques ensure mathematical equivalency between the synthesized net list and original RTL description. The net list contains no unmapped RTL statements and declarations.

The high-level synthesis tool performs the process of transforming the C-level models (SystemC, ANSI C/C++) description into a technology-dependent net list.

## Placement

The placement tool starts the physical implementation of the ASIC. With a 2-D floor plan provided by the ASIC designer, the placer tool assigns locations for each gate in the net list. The resulting placed gates net list contains the physical location of each of the net list's standard-cells, but retains an abstract description of how the gates' terminals are wired to each other.

Typically the standard cells have a constant size in at least one dimension that allows them to be lined up in rows on the integrated circuit. The chip will consist of a huge number of rows (with power and ground running next to each row) with each row filled with the various cells making up the actual design. Placers obey certain rules: Each gate is assigned a unique (exclusive) location on the die map. A given gate is placed once, and may not occupy or overlap the location of any other gate.

# Combinational Network Delay

The propagation of signals through a network is not instantaneous. This characteristic can be useful, for example, when creating circuits that output pulse signals. But it causes problems if the momentary changes of signals at the outputs lead to logical errors. Such transient output changes are called glitches. A logic circuit is said to have a hazard if it has the potential for these glitches.

Electronic signals travel along conductors at about 8cm per nanosecond (the actual speed depends on the conductor material, dimensions, and other external factors). Electronic switches, like the field-effect transistors (FETs) used in logic circuits, typically require up to several hundred picoseconds to turn on and off. When a switch does turn on, it must transfer charge to or from the capacitance at its output node, and again, this takes time. All of these factors contribute to the simple fact that time is required for electric signals to propagate through logic circuits. Restated, time is required to process information in digital circuits. This processing time is divided between the less significant signal transmission time, and the more significant propagation delays associated with switching logic circuits. If not managed properly, propagation delays can result in logic circuits that run too slowly to meet their requirements, or that fail altogether.

A simple logic circuit, its equivalent CMOS circuit, and a timing diagram are shown below in Figure below with a particular intra-gate node (N1) highlighted. The timing diagram illustrates logical behavior of signals as a function of time. Note that if B changes from low to high when C is high as shown, the circuit node N1 changes from high to low after a time $\tau_1$ has elapsed. The time $\tau_1$ is the "propagation delay" associated with the NAND gate. Referring to the CMOS circuit, the propagation delay τ1 models transistor Q1 turning on and discharging node N1 from Vdd to GND. Although there is no actual capacitor at the output node, all of the signal wires and FET connections associated with the circuit node N1 behave like a single capacitor, and this "parasitic" accumulated capacitance is shown lumped into a single component labeled C1. As is the case with any capacitor, C1 cannot transition from Vdd to GND immediately; the propagation delay τ1 models the time required to discharge this capacitance.

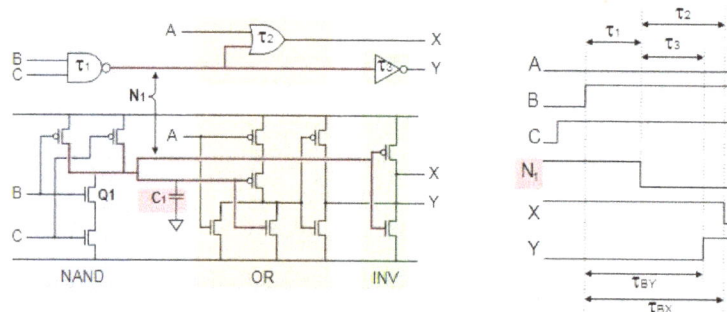

Figure: Propagation delay

As the C1 capacitance discharges, the voltage at N1 decreases below the input switching threshold of the inverter, the inverter drives its output Y to a '1' after the propagation delay τ3. The propagation delay of the OR gate (τ2) is longer than the delay for the inverter—in general, different gates will have different propagation delays. Further, since the delay through a given gate depends on the number of other gates and wires that it must drive, different instances of the same type of gate in a given circuit will have different propagation delays as well. In a given digital circuit, a designer is typically interested in the system response time rather than individual gate delays. For this circuit, the system response times TBX and TBY that show the time required for signals X and Y to change in response to a change on signal B are shown at the bottom of the timing diagram.

The amount of time required to drive an output from '0' to '1' (or vice-versa) depends on how much capacitance is present on the output node. In a CMOS circuit, the capacitance on a given output node is determined by how many "downstream" gate inputs are connected to the output node (for example, in the circuit above, node A is driving a single gate input while node N1 is driving two gate inputs). As a first approximation, it is reasonable to assume a linear relationship between the number of downstream gates driven by an output node and the amount of time required to transition the output node. That is, if an output node connected to 2 downstream gate inputs can transition from '0' to '1' in time X, the same gate driving 4 downstream gate inputs can transition in time 2X.

Different circuit implementation technologies have different typical delays. For example, a circuit implemented in a modern FPGA will typically have delays that are much smaller than a circuit implemented in a five-year-old FPGA, and in turn, both FPGA circuits would have far smaller delays than a similar circuit built from discrete gates. The smallest delay times (on the order of 10's of picoseconds) are available in the most expensive technologies, and these are reserved for "fully custom" chip designs that sell in high-volumes (like Pentium processors), or for designs that require the best performance for specialized applications (like sensitive scientific instruments). Whatever the technology, circuit delays are affected by variations in the manufacturing process, so no two devices from the same manufacturing line will exhibit exactly the same delay. Further, delays can change when circuits are exposed to different operating environments—both temperature and power supply voltage can greatly alter delays on various circuit nodes.

## Circuit Delays and CAD Tools

When a design is "implemented" (i.e., translated and mapped to a given technology) in a CAD tool like Xilinx's ISE/Webpack, a separate database containing specific information about every component in the design is created. This database contains information that defines the input/output relationships for each component, including the time required for input signal changes to propagate through the component to cause output signal changes. Delay information is typically stored separately for rising-edge transitions (i.e., a 0-to-1 transition) and for falling-edge transitions. Different delay values are used for rising and falling edges to account for the differences in the FETs that are used to drive an output node to '0' or '1'. In a falling transition, nFETs are responsible for driving the output node to '0', while in a rising transition, pFETs are responsible for driving an output node to '1'. In CMOS circuits, nFETs can typically pass twice the amount of current as similarly sized pFETs, so driving an output node to '1' typically takes twice as long as driving an output to '0'. Some simpler CAD tools ignore this phenomenon, and use a single number to define "gate delay". This single gate delay number is applied to all inputs for both rising and falling transitions.

In general, the delays encountered in a given circuit cannot be precisely known until the circuit is transformed into its most basic structural representation. The most basic representation depends on the technology that will be used to implement the circuit. When circuits are synthesized to a given device like an FPGA or CPLD, all the "logical" components and interconnections specified in the source file are mapped to particular physical devices in the chip. Once this mapping happens, it is possible to calculate the delays for every circuit node in the design with a high degree of accuracy. Prior to this mapping, it is only possible to estimate the delays. Whether calculated or estimated, all useful logic simulators must accommodate delay values so that designers can simulate the behavior of physical circuits. In fact, it is fair to say that accurate delay modeling is the most important and most useful feature of a simulator. Designers have learned that they must know the effects of all delays on all circuit nodes prior to releasing a design to manufacturing.

In a modern design flow, a circuit is initially designed without paying much attention to delays. In this early stage, a simulator is used only to check that the circuit logic has been correctly defined. When the design is synthesized to a given technology, the CAD tools can automatically calculate accurate delays for every single circuit node. Then, the circuit can be re-simulated, and the designer can study the circuit's behavior with accurate node delays included. Delay information is typically stored in a file called a "standard delay format", or .sdf file. In a post-synthesis simulation, the .sdf file is used by the simulator along with the circuit definition and the stimulus file to create a highly accurate output.

Many schematic-based CAD tools allow designers to include delays at the time a circuit is initially specified. These delays are by definition "best guesses", but they are nevertheless useful in studying a given circuit's performance. These delay values can easily be modified to simulate a circuit's behavior under different operating conditions that might arise. For example, best-case or worst-case delays could be used to model circuit performance at different operating temperatures or supply voltages.

## Timing Waveforms

Let's consider the circuit shown in figure below.

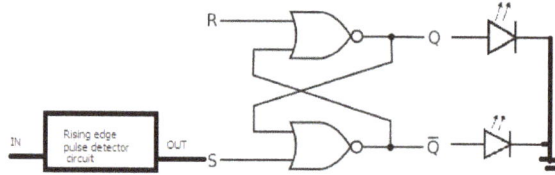

An input signal passes through three inversions, leaving it in its inverted state, which is then AND with the original input. This appears to implement a rather useless function: $A * A' = 0$ However, the timing diagram of figure above tells us a different story. After the input A goes high, the output waveform goes high for a short time before going low. Such a circuit is called a pulse shaper because a change at its input causes a short-duration pulse at the output.

The circuit of figure above operates as follows. Let's assume that the initial state has $A = 0$, $B = 1$, $C = 0$, $D = 1$, and $F = 0$, as shown in figure below at time step 0.

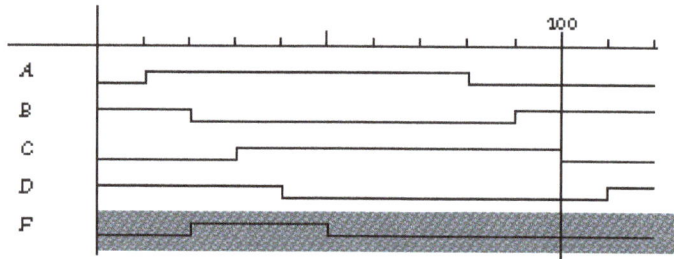

Further, we assume that each gate has a propagation delay of 10 time units. When input A changes from 0 to 1 at time 10, it takes 10 time units, a gate delay, before B changes from 1 to 0 (time step 20). After a second gate delay, C changes from 0 to 1 (time step 30). D changes from 1 to 0 after a third gate delay (time step 40). However, between time 10 and time 40, both A and D are logic 1. If the AND gate also has a 10-unit gate delay, the output F will be high between time steps 20 and 50. This is exactly what is shown in the timing diagram. In effect, the three inverters stretch the time during which A and D are both logic 1 after A changes from 0 to 1. Eventually, the change in A propagates to D as a 0, causing F to fall after another gate delay. It is no surprise that the pulse is exactly three inverter delays wide. If we increased the number of inverters to five, the width of the pulse would be five gate delays instead.

A pulse shaper circuit exploits the propagation asymmetries in signal paths with the explicit purpose of creating short-duration changes at the output. It generates a periodic waveform that could be used, for example, as a clock in a digital system. It operates much like a stopwatch. With its switch in one position, the circuit does nothing. In the second position, the circuit generates a periodic sequence of pulses.

## Analysis of a Pulse Shaper Circuit

In the analyzes of operation of the simple pulse shaper circuit of figure below.

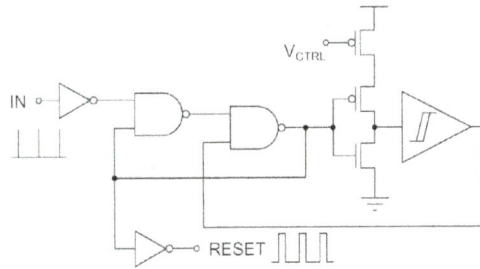

The circuit has a single input A that is connected to logic 1 when the switch is open and to logic 0 when the switch is closed. This is because the path to ground has lower resistance than the path to the power supply when the switch is closed. We will assume that the propagation delay of all gates is 10 time units.

Let's suppose that at time step 0, the switch has just been closed. We begin by determining the initial value for each of the circuit's nets. A goes to 0 instantly. Since a NAND gate will output a 1 whenever one of its inputs is 0, B goes to 1, but after a gate delay of 10 time units. So we say that B goes to 1 at time step 10.

C is set to the complement of B, but once again only after inverter propagation delay. Thus C goes to 0 at time step 20. D becomes the complement of C after another inverter delay. So it goes to 1 at time step 30. Since A is 0 and D is 1, the output of the NAND gate stays at 0. The circuit is said to be in a steady state.

What happens if the switch opens at time step 40? The input A immediately goes to 1. Now both inputs to the NAND gate are 1, so after a gate propagation of 10 time units, B will go low. This happens at time step 50.

The change in B propagates to C after another inverter delay. Thus at time step 60, C goes to 1. In a similar fashion, D goes to 0 at time step 70. Now the NAND gate has one of its inputs at 0, so at time step 80 B will go to 1.

Note that B first goes low at time step 50 and goes high at time step 80-a difference of 30 time units. This is exactly three gate delays: the delay through the NAND gate and the two inverter gates on the path from signal B to D.

Now that B is at 1, C will go to 0 at time step 90, D will go to 1 at time step 100, and B will return to 0 at time step 110. The circuit is no longer in steady state. It now oscillates with B, C, and D varying between 1 and 0, staying at each value for three gate delays (30 time units). The behavior of the circuit is summarized in the timing diagram.

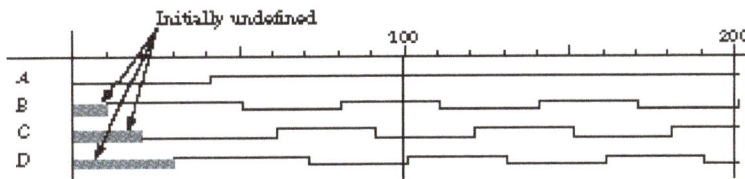

## References

- Combinatorial-logic: techtarget.com, Retrieved 28 March 2018

- Combinational-circuits, computer-logical-organization: tutorialspoint.com, Retrieved 31 March 2018

- Cmos-working-principle-and-applications: elprocus.com, Retrieved 08 May 2018

- Vlsi-design-digital-system: tutorialspoint.com, Retrieved 14 April 2018

- Low-Power-VLSI-Design, design-guide: eeherald.com, Retrieved 28 May 2018

- Design-for-manufacturability-29860: design-reuse.com, Retrieved 18 June 2018

# 3

# Sequential Machines

The field of electronics has undergone rapid developments in the past decades, which has resulted in innovation of electronic devices such as sequential machines. This chapter discusses the fundamentals of power optimization, clocking, flip flops, sequential systems and clock generators.

A sequential machine is a finite-state automaton with output (in some contexts including machines with infinite state set). Thus there is a function f from the Cartesian product $I \times Q$ to the product $Q \times O$, with $Q$ a set of states and $I$, $O$ finite sets of input and output symbols respectively. Suppose, for example, $a, q_0 \mapsto q_1, x b, q_1 \mapsto q_1, y c, q_1 \mapsto q_2, z$

Then, if the machine is in state $q_0$ and reads a, it moves to state $q_1$ and outputs $x$, and so on. Assuming the starting state to be $q_0$, it can be seen for example that the input string *abbbc* is mapped to the output string *xyyyz*. This mapping from the set of all input strings to the set of all output strings, i.e. $I^*$ to $O^*$, is called the response function of the machine. The function f comprises a state-transition function $f_Q$ from $I \times Q$ to $Q$ and an output function $f_O$ from $I \times Q$ to $O$.

What is described here is sometimes called a Mealy machine to distinguish it from the more restricted Moore machines. In a Moore machine, the symbol output at each stage depends only on the current state, and not on the input symbol read. The example above is therefore not a Moore machine since $f_O(b, q_1) = y$

whereas $f_O(c, q_1) = z$

Any Moore machine can be converted to an equivalent Mealy machine by adding more states.

A generalized sequential machine is an extension of the notion of sequential machine: a string of symbols is output at each stage rather than a single symbol. Thus there is a function from $I \times Q$ to $Q \times O^*$.

## Sequential Logic Circuits

Sequential Logic Circuits use flip-flops as memory elements and in which their output is dependent on the input state.

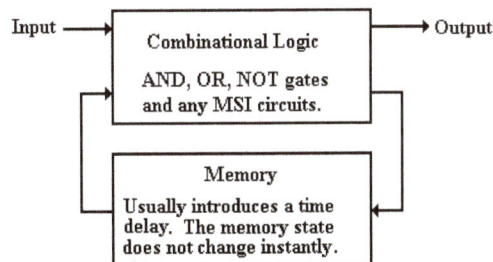

Unlike Combinational Logic circuits that change state depending upon the actual signals being applied to their inputs at that time, Sequential Logic circuits have some form of inherent "Memory" built in.

This means that sequential logic circuits are able to take into account their previous input state as well as those actually present, a sort of "before" and "after" effect is involved with sequential circuits.

In other words, the output state of a "sequential logic circuit" is a function of the following three states, the "present input", the "past input" and the "past output". Sequential Logic circuits remember these conditions and stay fixed in their current state until the next clock signal changes one of the states, giving sequential logic circuits "Memory".

Sequential logic circuits are generally termed as two state or Bistable devices which can have their output or outputs set in one of two basic states, a logic level "1" or a logic level "0" and will remain "latched" (hence the name latch) indefinitely in this current state or condition until some other input trigger pulse or signal is applied which will cause the bistable to change its state once again.

## Sequential Logic Representation

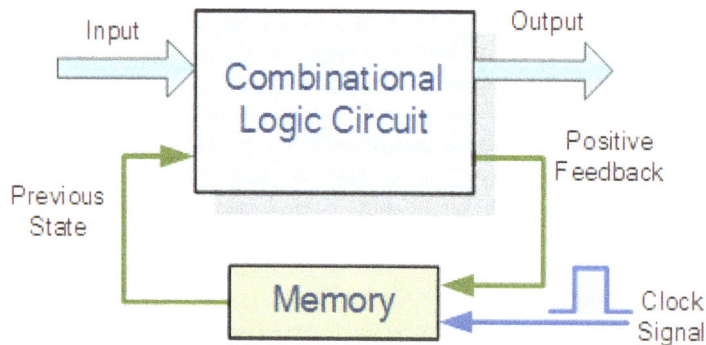

The word "Sequential" means that things happen in a "sequence", one after another and in Sequential Logic circuits, the actual clock signal determines when things will happen next. Simple sequential logic circuits can be constructed from standard Bistable circuits such as: Flip-flops, Latches and Counters and which themselves can be made by simply connecting together universal NAND Gates and NOR Gates in a particular combinational way to produce the required sequential circuit.

## Classification of Sequential Logic

As standard logic gates are the building blocks of combinational circuits, bistable latches and flip-flops are the basic building blocks of sequential logic circuits. Sequential logic circuits can be constructed to produce either simple edge-triggered flip-flops or more complex sequential circuits such as storage registers, shift registers, memory devices or counters. Either way sequential logic circuits can be divided into the following three main categories:

- Event Driven: asynchronous circuits that change state immediately when enabled.

- Clock Driven: synchronous circuits that are synchronized to a specific clock signal.

- Pulse Driven: which is a combination of the two that responds to triggering pulses.

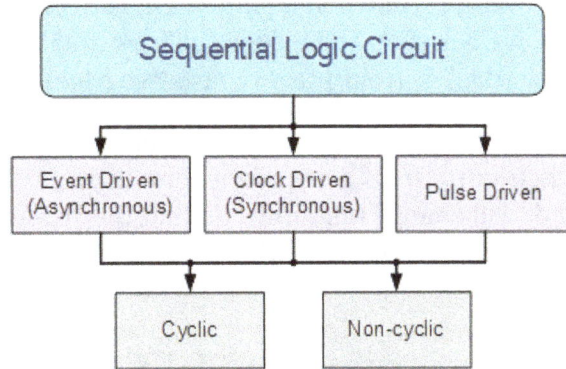

As well as the two logic states mentioned above logic level "1" and logic level "0", a third element is introduced that separates sequential logic circuits from their combinational logic counterparts, namely TIME. Sequential logic circuits return back to their original steady state once reset and sequential circuits with loops or feedback paths are said to be "cyclic" in nature.

We now know that in sequential circuits changes occur only on the application of a clock signal making it synchronous; otherwise the circuit is asynchronous and depends upon an external input. To retain their current state, sequential circuits rely on feedback and this occurs when a fraction of the output is fed back to the input and this is demonstrated as:

## Sequential Feedback Loop

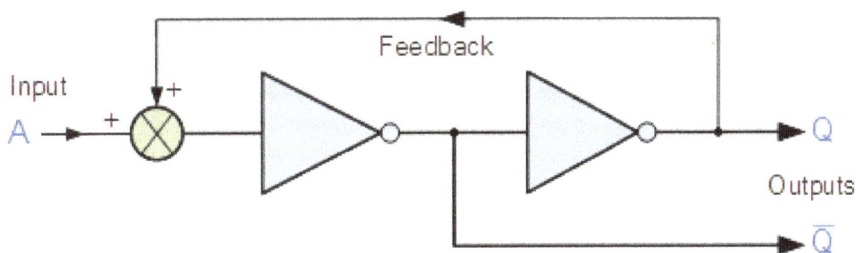

The two inverters or NOT gates are connected in series with the output at Q fed back to the input. Unfortunately, this configuration never changes state because the output will always be the same, either a "1" or a "0", it is permanently set.

## Importance of Clock Signal in Sequential Circuits

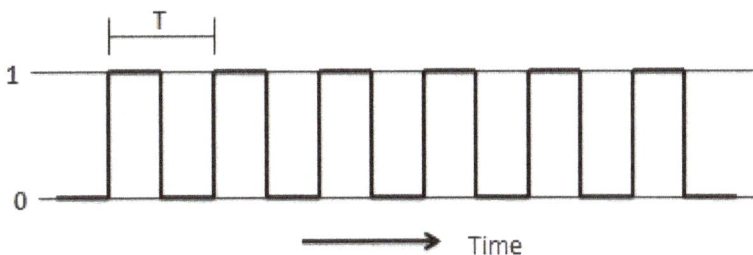

The clock signal plays a crucial role in sequential circuits. A clock is a signal, which oscillates between logic level 0 and logic level 1, repeatedly. Square wave with constant frequency is the most common form of clock signal. A clock signal has "edges". These are the instants at which the clock changes from 0 to 1 (a positive edge) or from 1 to 0 (a negative edge).

Clock signals control the outputs of the sequential circuit .That is it determines when and how the memory elements change their outputs. If a sequential circuit is not having any clock signal as input, the output of the circuit will change randomly. So that they cannot retain their state till the next input signal arrives. But sequential circuits with clock input will retain its state till the next clock edge occurs.

## Classification of Sequential Circuits on the Basis of Clock Signal

Based on the clock signal input, the sequential circuits are classified into two types.

- Synchronous sequential circuit
- Asynchronous sequential circuit

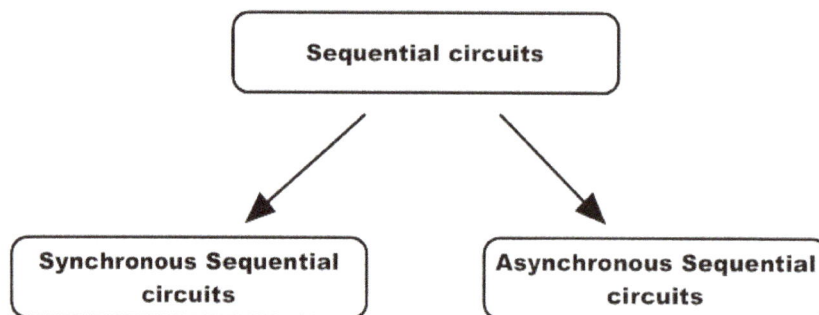

## Synchronous Sequential Circuits

In Synchronous sequential circuit, the output depends on present and previous states of the inputs at the clocked instances. The circuits use a memory element to store the previous state. The memory elements in these circuits will have clocks. All these clock signals are driven by the same clock signal.

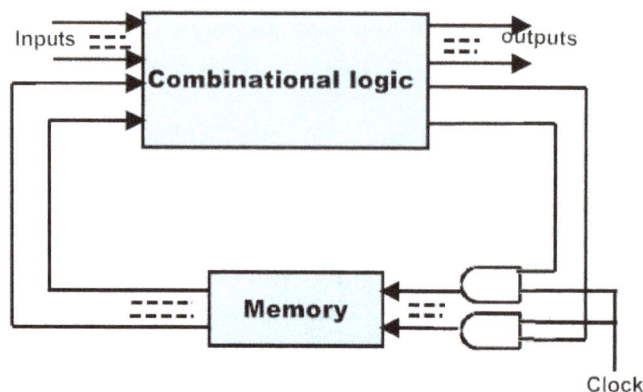

- Using clock signal, state changes will occur across all storage elements.

- These circuits are bit slower compared to asynchronous because they wait for the next clock pulse to arrive to perform the next operation.

- These circuits can be clocked or pulsed.

- The Synchronous sequential circuits that use clock pulses in their inputs are called clocked-sequential circuits. They are very stable.

- The sequential circuits that change their state using the pulse and these are called pulsed or unclocked sequential circuits.

## Places Where Synchronous Sequential Circuits are used

- Used in the design of Moore-Mealy state management machines.

- They are used in synchronous counters, flip flops etc.

## Limitations of Synchronous Sequential Circuits

- All the flip – flops in synchronous sequential circuits must be connected to clock signal. Clock signals are very high frequency signals and clock distribution consumes and dissipated a large amount of heat.

- Critical path or the slowest path determines the maximum possible clock frequency. Hence they are slower than asynchronous circuits.

## Asynchronous Sequential Circuits

The Sequential circuits which do not operate by clock signals are called "Asynchronous sequential circuits".

- These circuits will change their state immediately when there is a change in the input signal.

- The Circuit behavior is determined by signals at any instant in time and the order in which input signals change.

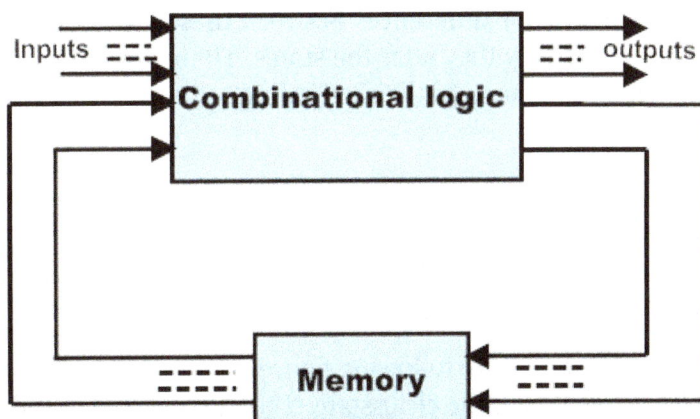

## They Do not Operate in Pulse Mode

- They have better performance but hard to design due to timing problems.

- Mostly we use the asynchronous circuits when we require the low power operations.

- They are faster than synchronous sequential circuits as they do not need to wait for any clock signal.

## Places Where Asynchronous Sequential Circuits are used

These are used when speed of operation is important. As they are independent of internal clock pulse, they are operate quickly. So they are used in Quick response circuits.

Used in the communication between two units having their own independent clocks.

Used when we require the better external input handling.

## Drawbacks

- Asynchronous sequential circuits are more difficult to design.

- Though they have a faster performance, their output is uncertain.

# Flip Flop

The term flip-flop (FF) was invented in the year 1918 by the British physicist F.W Jordan and William Eccles. It was named as the Eccles Jordan trigger circuit and includes two active elements. The design of the FF was used in British Colossus code breaking computer the year 1943. The transistorized versions of these circuits were common in computers, even after the overview of integrated circuits, though FFs made from logic gates are also common now. The first flip-flop circuit were known differently as multi vibrators or trigger circuits.

FF is a circuit element where the o/p not only depends on the present inputs, but also depends on the former input and o/ps. The major difference between flip flop circuit and a latch is that a FF includes a clock signal, whereas a latch doesn't. Basically, there are four kinds of latches & FFs namely: T, D, SR and JK. The major differences between these kinds of FFs and latches are the number of inputs they have and how they alter the states. There are different differences for each kind of FFs and latches which can increase their operations.

## Flip Flop Circuit

The designing of the flip flop circuit can be done by using logic gates such as two NAND and NOR gates. Each flip flop consists of two inputs and two outputs, namely set and reset, Q and Q'. This kind of flip flop is stated to as an SR flip flop or SR latch.

The FF includes two states shown in the following figure. When Q=1 andQ'=0 then it is in the set state. When Q=0 and Q'=1then it is in the clear state. The FF's outputs Q and Q' are complements of each other and that are stated to as the normal & complement outputs respectively. The binary state of the flip flop is taken to be the normal output value.

When the input 1 is applied to the flip flop, both the outputs of the FF go to 0, so both the o/p's are complements of each other. In a regular operation, this ailment must be neglected by making sure that ones are not applied to both the inputs concurrently.

## Types of Flip-flops

Flip flop circuits are classified into four types based on its use, namely D-Flip Flop, T- Flip Flop, SR- Flip Flop and JK- Flip Flop.

## SR Flip-flop

The SR flip-flop, also known as a SR Latch, can be considered as one of the most basic sequential logic circuit possible. This simple flip-flop is basically a one-bit memory bistable device that has two inputs, one which will "SET" the device (meaning the output = "1"), and is labeled S and one which will "RESET" the device (meaning the output = "0"), labeled R.

Then the SR description stands for "Set-Reset". The reset input resets the flip-flop back to its original state with an output Q that will be either at a logic level "1" or logic "0" depending upon this set/reset condition.

A basic NAND gate SR flip-flop circuit provides feedback from both of its outputs back to its opposing inputs and is commonly used in memory circuits to store a single data bit. Then the SR flip-flop actually has three inputs, Set, Reset and its current output Q relating to its current state or history. The term "Flip-flop" relates to the actual operation of the device, as it can be "flipped" into one logic Set state or "flopped" back into the opposing logic Reset state.

## NAND Gate SR Flip-flop

The simplest way to make any basic single bit set-reset SR flip-flop is to connect together a pair of cross-coupled 2-input NAND gates as shown, to form a Set-Reset Bistable also known as an active LOW SR NAND Gate Latch, so that there is feedback from each output to one of the other NAND gate inputs. This device consists of two inputs, one called the Set, S and the other called the Reset, R with two corresponding outputs Q and its inverse or complement $\overline{Q}$ (not-Q) as shown below.

## Basic SR Flip-flop

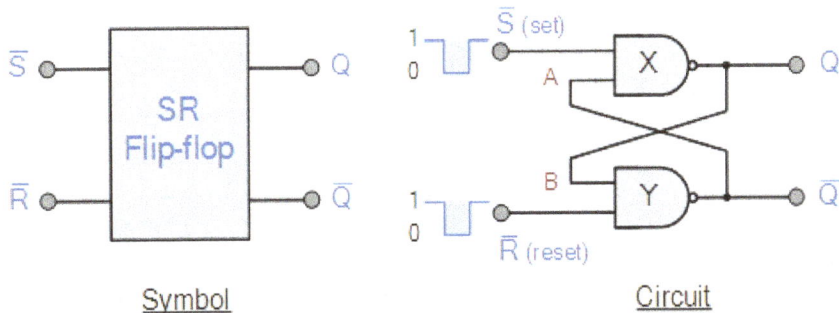

Symbol     Circuit

## Set State

Consider the circuit shown above. If the input R is at logic level "0" (R = 0) and input S is at logic level "1" (S = 1), the NAND gate Y has at least one of its inputs at logic "0" therefore, its output $\overline{Q}$ must be at a logic level "1" (NAND Gate principles). Output $\overline{Q}$ is also fed back to input "A" and so both inputs to NAND gate X are at logic level "1", and therefore its output Q must be at logic level "0".

Again NAND gate principals. If the reset input R changes state, and goes HIGH to logic "1" with S remaining HIGH also at logic level "1", NAND gate Y inputs are now R = "1" and B = "0". Since one of its inputs is still at logic level "0" the output at $\overline{Q}$ still remains HIGH at logic level "1" and there is no change of state. Therefore, the flip-flop circuit is said to be "Latched" or "Set" with $\overline{Q}$ = "1" and Q = "0".

## Reset State

In this second stable state, $\overline{Q}$ is at logic level "0", (not Q = "0") its inverse output at Q is at logic level "1", (Q = "1"), and is given by R = "1" and S = "0". As gate X has one of its inputs at logic "0" its output Q must equal logic level "1" (again NAND gate principles). Output Q is fed back to input "B", so both inputs to NAND gate Y are at logic "1", therefore, $^{—}$ = "0".

If the set input, S now changes state to logic "1" with input R remaining at logic "1", output $\overline{Q}$ still remains LOW at logic level "0" and there is no change of state. Therefore, the flip-flop circuits "Reset" state has also been latched and we can define this "set/reset" action in the following truth table.

## Truth Table for this Set-reset Function

| State | S | R | Q | Q | Description |
|-------|---|---|---|---|-------------|
| Set | 1 | 0 | 0 | 1 | Set Q » 1 |
|  | 1 | 1 | 0 | 1 | no change |
| Reset | 0 | 1 | 1 | 0 | Reset Q » 0 |
|  | 1 | 1 | 1 | 0 | no change |
| Invalid | 0 | 0 | 1 | 1 | Invalid Condition |

It can be seen that when both inputs S = "1" and R = "1" the outputs Q and $\overline{Q}$ can be at either logic level "1" or "0", depending upon the state of the inputs S or R BEFORE this input condition existed. Therefore the condition of S = R = "1" does not change the state of the outputs Q and $\overline{Q}$.

However, the input state of S = "0" and R = "0" is an undesirable or invalid condition and must be avoided. The condition of S = R = "0" causes both outputs Q and $\overline{Q}$ to be HIGH together at logic level "1" when we would normally want $\overline{Q}$ to be the inverse of Q. The result is that the flip-flop loses control of Q and $\overline{Q}$, and if the two inputs are now switched "HIGH" again after this condition to logic "1", the flip-flop becomes unstable and switches to an unknown data state based upon the unbalance as shown in the following switching diagram.

## S-R Flip-flop Switching Diagram

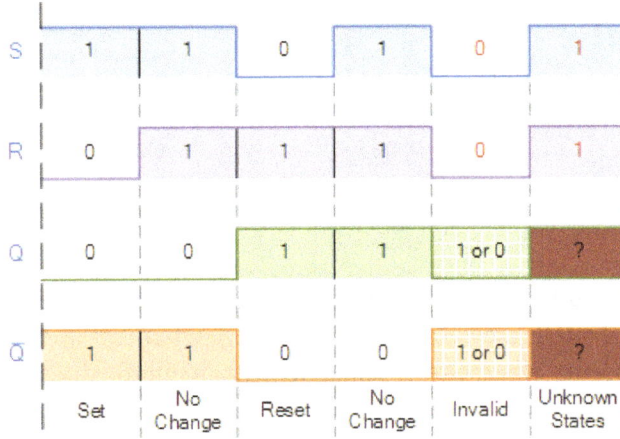

This unbalance can cause one of the outputs to switch faster than the other resulting in the flip-flop switching to one state or the other which may not be the required state and data corruption will exist. This unstable condition is generally known as its Meta-stable state.

Then, a simple NAND gate SR flip-flop or NAND gate SR latch can be set by applying a logic "0", (LOW) condition to its Set input and reset again by then applying a logic "0" to its Reset input. The SR flip-flop is said to be in an "invalid" condition (Meta-stable) if both the set and reset inputs are activated simultaneously.

As we have seen above, the basic NAND gate SR flip-flop requires logic "0" inputs to flip or change state from Q to $\overline{Q}$ and vice versa. We can however, change this basic flip-flop circuit to one that changes state by the application of positive going input signals with the addition of two extra NAND gates connected as inverters to the S and R inputs as shown.

## Positive NAND Gate SR Flip-flop

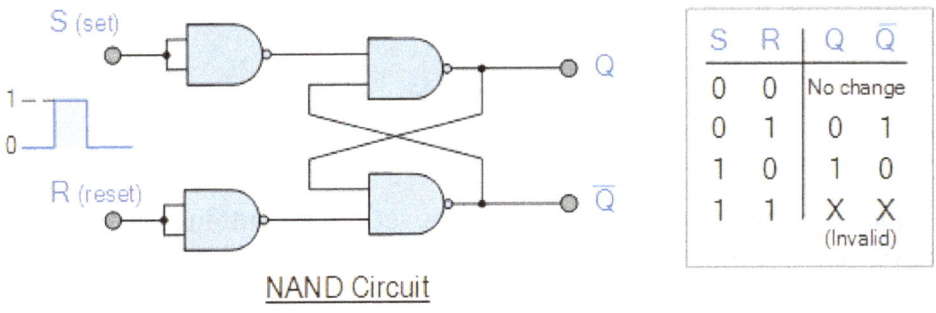

NAND Circuit

| S | R | Q | $\overline{Q}$ |
|---|---|---|---|
| 0 | 0 | No change | |
| 0 | 1 | 0 | 1 |
| 1 | 0 | 1 | 0 |
| 1 | 1 | X | X |
| | | (Invalid) | |

As well as using NAND gates, it is also possible to construct simple one-bit SR Flip-flops using two cross-coupled NOR gates connected in the same configuration. The circuit will work in a similar way to the NAND gate circuit above, except that the inputs are active HIGH and the invalid condition exists when both its inputs are at logic level "1", and this is shown below.

## NOR Gate SR Flip-flop

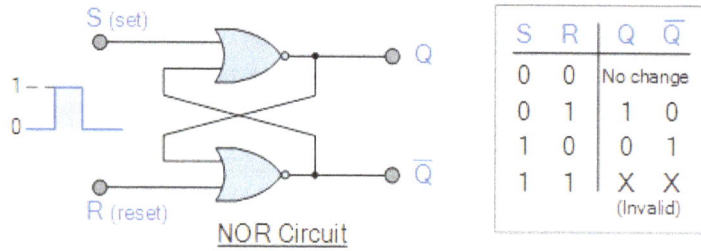

| S | R | Q | $\overline{Q}$ |
|---|---|---|---|
| 0 | 0 | No change | |
| 0 | 1 | 1 | 0 |
| 1 | 0 | 0 | 1 |
| 1 | 1 | X | X |
| | | (Invalid) | |

NOR Circuit

## Switch Debounce Circuits

Edge-triggered flip-flops require a nice clean signal transition, and one practical use of this type of set-reset circuit is as a latch used to help eliminate mechanical switch "bounce". As its name implies, switch bounce occurs when the contacts of any mechanically operated switch, push-button or keypad are operated and the internal switch contacts do not fully close cleanly, but bounce together first before closing (or opening) when the switch is pressed.

This gives rise to a series of individual pulses which can be as long as tens of milliseconds that an electronic system or circuit such as a digital counter may see as a series of logic pulses instead of one long single pulse and behave incorrectly. For example, during this bounce period the output voltage can fluctuate wildly and may register multiple input counts instead of one single count. Then set-reset SR Flip-flops or Bistable Latch circuits can be used to eliminate this kind of problem and this is demonstrated below.

## SR Flip-flop Switch Debounce Circuit

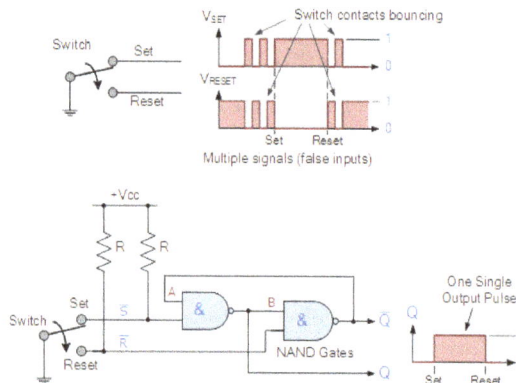

Depending upon the current state of the output, if the set or reset buttons are depressed the output will change over in the manner described above and any additional unwanted inputs (bounces) from the mechanical action of the switch will have no effect on the output at Q.

When the other button is pressed, the very first contact will cause the latch to change state, but any additional mechanical switch bounces will also have no effect. The SR flip-flop can then be RESET automatically after a short period of time, for example 0.5 seconds, so as to register any additional and intentional repeat inputs from the same switch contacts, such as multiple inputs from a keyboards "RETURN" key.

Commonly available IC's specifically made to overcome the problem of switch bounce are the MAX6816, single input, MAX6817, dual input and the MAX6818 octal input switch debouncer IC's. These chips contain the necessary flip-flop circuitry to provide clean interfacing of mechanical switches to digital systems.

Set-Reset bistable latches can also be used as Monostable (one-shot) pulse generators to generate a single output pulse, either high or low, of some specified width or time period for timing or control purposes. The 74LS279 is a Quad SR Bistable Latch IC, which contains four individual NAND type bistable's within a single chip enabling switch debounce or monostable/astable clock circuits to be easily constructed.

## Quad SR Bistable Latch 74LS279

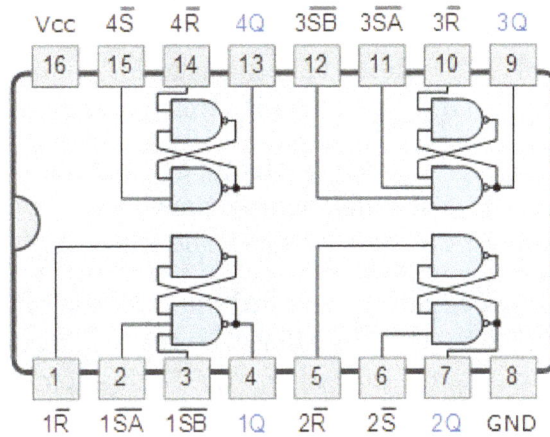

## Gated or Clocked SR Flip-flop

It is sometimes desirable in sequential logic circuits to have a bistable SR flip-flop that only changes state when certain conditions are met regardless of the condition of either the Set or the Reset inputs. By connecting a 2-input AND gate in series with each input terminal of the SR Flip-flop a Gated SR Flip-flop can be created. This extra conditional input is called an "Enable" input and is given the prefix of "EN". The addition of this input means that the output at Q only changes state when it is HIGH and can therefore be used as a clock (CLK) input making it level-sensitive as shown below.

## Gated SR Flip-flop

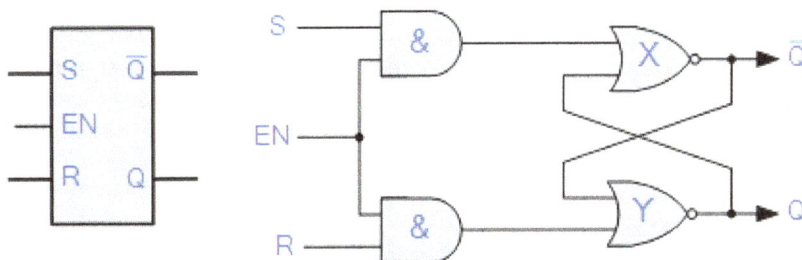

When the Enable input "EN" is at logic level "0", the outputs of the two AND gates are also at logic level "0", (AND Gate principles) regardless of the condition of the two inputs S and R, latching the two outputs Q and $\overline{Q}$ into their last known state. When the enable input "EN" changes to logic level "1" the circuit responds as a normal SR bistable flip-flop with the two AND gates becoming transparent to the Set and Reset signals.

This additional enable input can also be connected to a clock timing signal (CLK) adding clock synchronization to the flip-flop creating what is sometimes called a "Clocked SR Flip-flop". So a Gated Bistable SR Flip-flop operates as a standard bistable latch but the outputs are only activated when a logic "1" is applied to its EN input and deactivated by a logic "0".

## JK Flip-flop

JK flip Flop is the most widely used of all the flip-flop designs and is considered to be a universal flip-flop circuit. The two inputs labeled "J" and "K" are not shortened abbreviated letters of other words, such as "S" for Set and "R" for Reset, but are themselves autonomous letters chosen by its inventor Jack Kirby to distinguish the flip-flop design from other types.

The sequential operation of the JK flip flop is exactly the same as for the previous SR flip-flop with the same "Set" and "Reset" inputs. The difference this time is that the "JK flip flop" has no invalid or forbidden input states of the SR Latch even when S and R are both at logic "1".

The JK flip flop is basically a gated SR flip-flop with the addition of a clock input circuitry that prevents the illegal or invalid output condition that can occur when both inputs S and R are equal to logic level "1". Due to this additional clocked input, a JK flip-flop has four possible input combinations, "logic 1", "logic 0", "no change" and "toggle". The symbol for a JK flip flop is similar to that of an SR Bistable Latch.

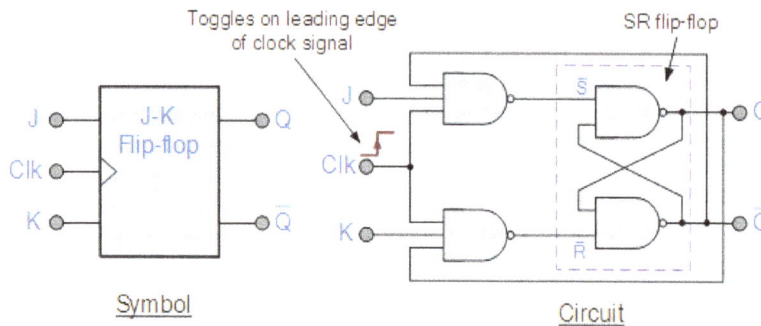

Symbol                                    Circuit

Both the S and the R inputs of the previous SR bistable have now been replaced by two inputs called the J and K inputs, respectively after its inventor Jack Kilby. Then this equates to: J = S and K = R.

The two 2-input AND gates of the gated SR bistable have now been replaced by two 3-input NAND gates with the third input of each gate connected to the outputs at Q and $\overline{Q}$. This cross coupling of the SR flip-flop allows the previously invalid condition of S = "1" and R = "1" state to be used to produce a "toggle action" as the two inputs are now interlocked.

If the circuit is now "SET" the J input is inhibited by the "0" status of $\overline{Q}$ through the lower NAND gate. If the circuit is "RESET" the K input is inhibited by the "0" status of Q through the upper NAND gate. As Q and $\overline{Q}$ are always different we can use them to control the input. When both inputs J and K are equal to logic "1", the JK flip flop toggles as shown in the following truth table.

## Truth Table for the JK Function

| same as | Input | | Output | | Description |
|---|---|---|---|---|---|
| for the | J | K | Q | Q | |
| SR Latch | 0 | 0 | 0 | 0 | Memory |
| | 0 | 0 | 0 | 1 | no change |
| | 0 | 1 | 1 | 0 | Reset Q » 0 |
| | 0 | 1 | 0 | 1 | |
| | 1 | 0 | 0 | 1 | Set Q » 1 |
| | 1 | 0 | 1 | 0 | |
| Toggle | 1 | 1 | 0 | 1 | Toggle |
| action | 1 | 1 | 1 | 0 | |

Then the JK flip-flop is basically an SR flip flop with feedback which enables only one of its two input terminals, either SET or RESET to be active at any one time thereby eliminating the invalid condition seen previously in the SR flip flop circuit.

Also when both the J and the K inputs are at logic level "1" at the same time, and the clock input is pulsed "HIGH", the circuit will "toggle" from its SET state to a RESET state, or vi-sa-versa. This results in the JK flip flop acting more like a T-type toggle flip-flop when both terminals are "HIGH".

Although this circuit is an improvement on the clocked SR flip-flop it still suffers from timing problems called "race" if the output Q changes state before the timing pulse of the clock input has time to go "OFF". To avoid this the timing pulse period (T) must be kept as short as possible (high frequency). As this is sometimes not possible with modern TTL IC's the much improved Master-Slave JK Flip-flop was developed.

## Master-slave JK Flip-flop

The master-slave flip-flop eliminates all the timing problems by using two SR flip-flops connected together in a series configuration. One flip-flop acts as the "Master" circuit, which triggers on the leading edge of the clock pulse while the other acts as the "Slave" circuit, which triggers on the falling edge of the clock pulse. This results in the two sections, the master section and the slave section being enabled during opposite half-cycles of the clock signal.

The TTL 74LS73 is a Dual JK flip-flop IC, which contains two individual JK type bistable's within a single chip enabling single or master-slave toggle flip-flops to be made. Other JK flip flop IC's include the 74LS107 Dual JK flip-flop with clear, the 74LS109 Dual positive-edge triggered JK flip flop and the 74LS112 Dual negative-edge triggered flip-flop with both preset and clear inputs.

## Dual JK Flip-flop 74LS73

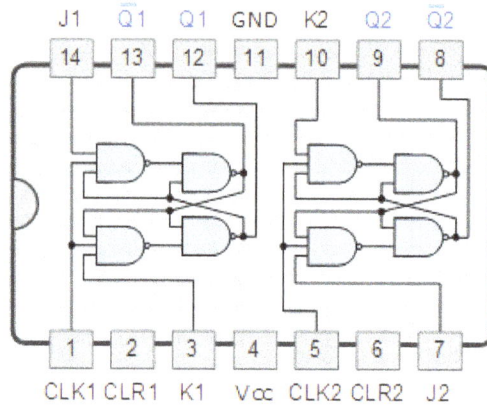

## Other Popular JK Flip-flop ICs

| Device Number | Subfamily | Device Description |
|---|---|---|
| 74LS73 | LS TTL | Dual JK-type Flip Flops with Clear |
| 74LS76 | LS TTL | Dual JK-type Flip Flops with Preset and Clear |
| 74LS107 | LS TTL | Dual JK-type Flip Flops with Clear |
| 4027B | Standard CMOS | Dual JK-type Flip Flop |

## Master-slave JK Flip-flop

The Master-slave Flip-flop is basically two gated SR flip-flops connected together in a series configuration with the slave having an inverted clock pulse. The outputs from Q and $\overline{Q}$ from the "Slave" flip-flop are fed back to the inputs of the "Master" with the outputs of the "Master" flip flop being connected to the two inputs of the "Slave" flip flop. This feedback configuration from the slave's output to the master's input gives the characteristic toggle of the JK flip flop as shown below.

## Master-slave JK Flip Flop

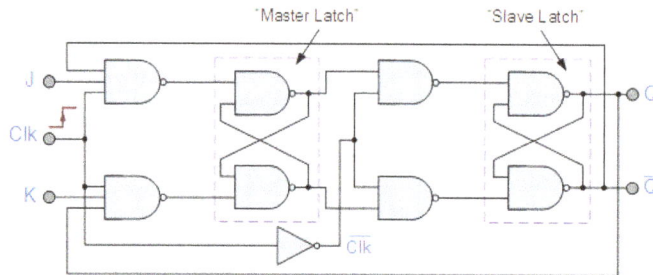

The input signals J and K are connected to the gated "master" SR flip flop which "locks" the input condition while the clock (Clk) input is "HIGH" at logic level "1". As the clock input of the "slave" flip flop is the inverse (complement) of the "master" clock input, the "slave" SR flip flop does not toggle. The outputs from the "master" flip flop are only "seen" by the gated "slave" flip flop when the clock input goes "LOW" to logic level "o".

When the clock is "LOW", the outputs from the "master" flip flop are latched and any additional changes to its inputs are ignored. The gated "slave" flip flop now responds to the state of its inputs passed over by the "master" section.

Then on the "Low-to-High" transition of the clock pulse the inputs of the "master" flip flop are fed through to the gated inputs of the "slave" flip flop and on the "High-to-Low" transition the same inputs are reflected on the output of the "slave" making this type of flip flop edge or pulse-triggered.

Then, the circuit accepts input data when the clock signal is "HIGH", and passes the data to the output on the falling-edge of the clock signal. In other words, the Master-Slave JK Flip flop is a "Synchronous" device as it only passes data with the timing of the clock signal.

## D Flip-flop

A D type (Data or delay flip flop) has a single data input in addition to the clock input as shown in figure below.

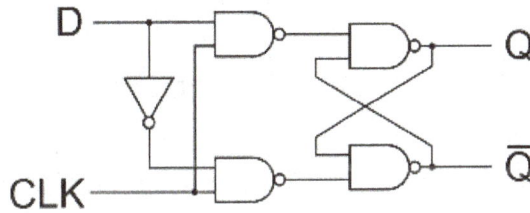

Figure: D Flip Flop

Basically, such type of flip flop is a modification of clocked RS flip flop gates from a basic Latch flip flop and NOR gates modify it in to a clock RS flip flop. The D input goes directly to S input and its complement through NOT gate is applied to the R input.

This kind of flip flop prevents the value of D from reaching the output until a clock pulse occurs. The action of circuit is straight forward as follows.

When the clock is low, both AND gates are disabled, therefore D can change values without affecting the value of Q. On the other hand, when the clock is high, both AND gates are enabled. In this case, Q is forced equal to D when the clock again goes low, Q retains or stores the last value of D. The truth table for such a flip flop is as given below in table below.

Table: Truth table for D Flip Flop

| S | R | Q(t + 1) |
|---|---|---|
| 0 | 0 | 0 |
| 0 | 1 | 1 |
| 1 | 0 | 0 |
| 1 | 1 | 1 |

Table: Excitation table for D Flip Flop

| S | Q |
|---|---|
| 0 | 0 |
| 1 | 1 |

## T Flip- lop

A method of avoiding the indeterminate state found in the working of RS flip flop is to provide only one input (the T input) such, flip flop acts as a toggle switch. Toggle means to change in the previous stage i.e. switch to opposite state. It can be constructed from clocked RS flip flop be incorporating feedback from output to input as shown in figure below.

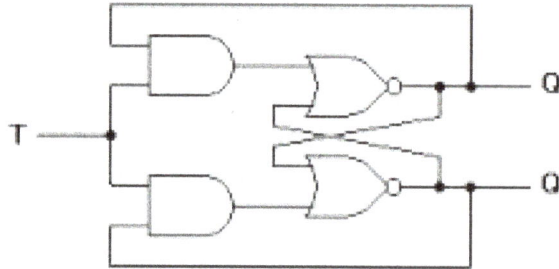

Figure: T Flip Flop

Such a flip flop is also called toggle flip flop. In such a flip flop a train of extremely narrow triggers drives the T input each time one of these triggers, the output of the flip flop changes stage. For in-stance Q equals o just before the trigger. Then the upper AND gate is enable and the lower AND gate is disabled. When the trigger arrives, it results in a high S input.

This sets the Q output to 1. When the next trigger appears at the point T, the lower AND gate is en-abled and the trigger passes through to the R input this forces the flip flop to reset.

Since each incoming trigger is alternately changed into the set and reset inputs the flip flop toggles. It takes two triggers to produce one cycle of the output waveform. This means the output has half the frequency of the input stated another way, a T flip flop divides the input frequency by two. Thus such a circuit is also called a divide by two circuits.

A disadvantage of the toggle flip flop is that the state of the flip flop after a trigger pulse has been applied is only known if the previous state is known. The truth table for a T flip flop is as given table below:

| $Q_n$ | T | $Q_n + 1$ |
|---|---|---|
| 0 | 0 | 0 |
| 0 | 1 | 1 |
| 1 | 0 | 1 |
| 1 | 1 | 0 |

Table: Truth table for T Flip Flop

The excitation table for T flip flop is very simply derived as shown in Table below.

| T | Q |
|---|---|
| 0 | $Q_n$ |
| 1 | n |

Table: Excitation table for T Flip Flop

# Sequential Systems

Sequential logic systems involve feedback, so that the previous state of the output of the logic system also has an impact on whether a change in the inputs produces a change in the output(s).

## D-type Flip-flops

There are many uses for D-type flip-flops in electronics such as:

1. Data transfer

2. Latches

3. Counters

D-type flip-flops are available in a 14-pin dual-in-line (DIL) package, as they contain two identical but completely separate D-type flip-flop circuits. The symbol for this D-type flip-flop is shown opposite.

The terminal labeled 'D' is called the data input terminal, and this is where the flip-flop receives data.

The terminal labeled 'Q' is the output terminal, and $\overline{Q}$ is the inverse output terminal where the logic level will always be the opposite of 'Q'.

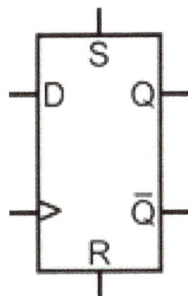

The terminal labeled '>' is called the clock input.

Two additional connections are also shown, S and R. These are connections which enable the user to SET the output (make Q = 1, and $\overline{Q}$ = 0) by applying a logic 1 to the S input. Similarly the user can RESET the output (make Q = 0, $\overline{Q}$ = 1) by applying a logic 1 to the R input

Q will remain at logic 0 for as long as R is at logic 1. These changes happen irrespective of the state of either D or the clock (>) inputs.

The R and S inputs will automatically appear in simulation circuits.

The R input will only be shown in circuits that require its use. We will not consider the S input in this course and it will not be shown on these diagrams.

## Data Transfer

When a logic 1 signal is applied to a clock (>) input, whatever logic state is present at the D, input will be transferred to the Q output. Special circuitry inside the D-type ensures that this transfer only occurs when the clock signal is changing from logic 0 and logic 1.

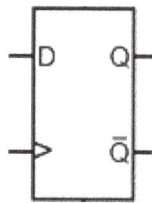

This action is usually referred to as rising-edge triggered. Once the clock signal reaches logic 1 any further changes at the D input will not be transferred to the output until another rising edge signal is applied to the clock input.

## Example

The diagram opposite shows a rising-edge-triggered D-type flip-flop.

The following graphs show the signals applied to the D and clock (>) inputs. Complete the remaining graphs to show the output Q and $\overline{Q}$

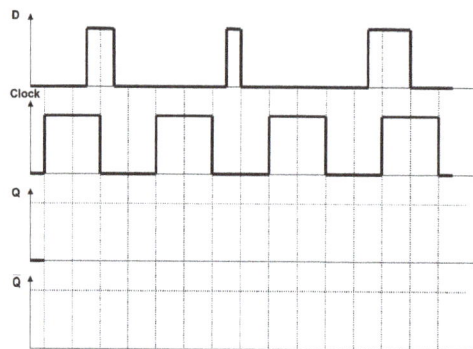

In this example, only the rising edges of the clock pulses are important, since this is the only time that the logic state of D can be transferred to Q.

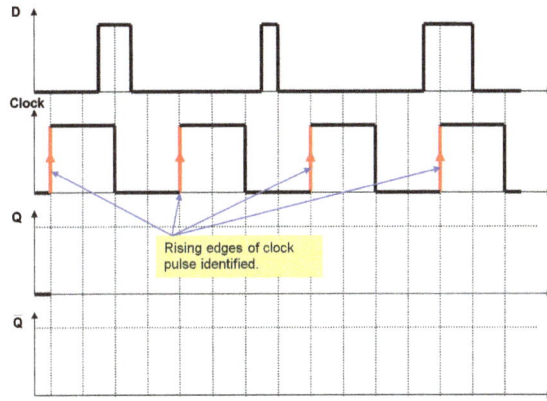

Step 1: identify the rising edges of the clock pulses

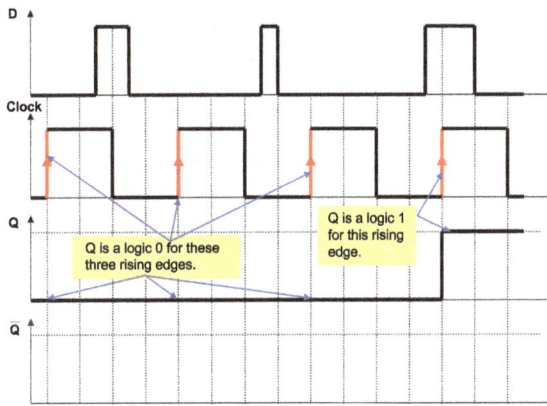

Step 2: transfer the logic state of D, to Q only at the times where the clock pulse is rising.

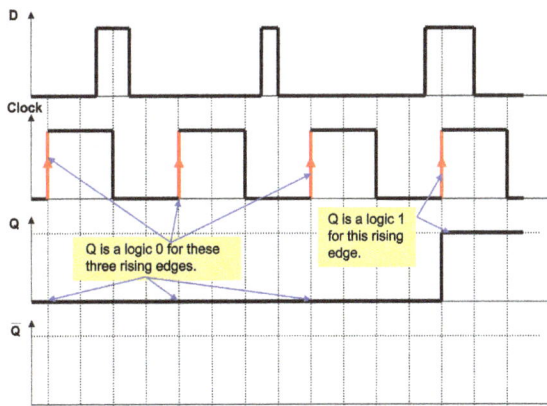

Step 3: Complete $\overline{Q}$ which will be opposite of Q.

## D-type Flip-flop Used as a Latch

As well as transferring data from one terminal to another on the rising edge of the clock pulse a D-type flip-flop can also be used to make a latch.

The circuit diagram for this application is shown below:

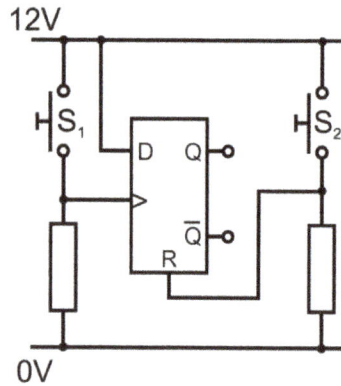

Notice that the D input is permanently connected to the positive of the power supply (logic level 1). A momentary press on switch S1 provides a rising-edge clock pulse to the D-type and this logic 1 from the D input is passed through to the Q output. The Q output will remain at logic 1 until the D-type is reset by momentarily pressing switch S2.

The action of the latch is summarized in the following graph:

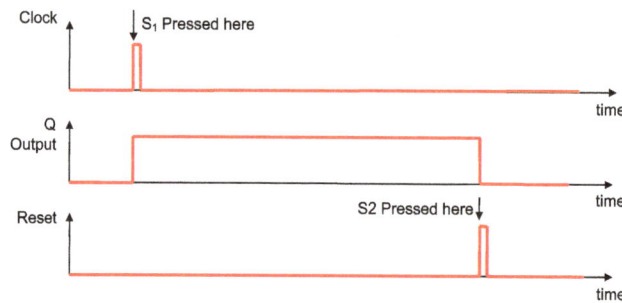

The rising-edge clock pulse could also be provided by a sensing sub-system or the output from a logic gate.

## Binary Counters

Often we need to count events, such as the number of boxes moving along a conveyor belt or the number of cars entering a car park. We can use electronic counters to perform the counting.

## Configuring the D-Type Flip-flop to Produce a Divide-by-two Function

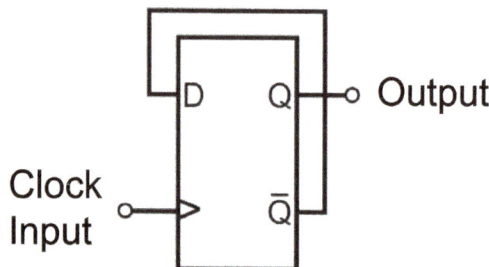

Remember: changes occur only on the rising edge of the clock pulse, and then the value of D is copied to Q.

Notice that the only connection is the link between $\overline{Q}$ and D.

A pulse generator is connected to the clock input.

Initially, Q and clock are at logic 0, and $\overline{Q}$ and D at logic 1. The timing diagram follows:

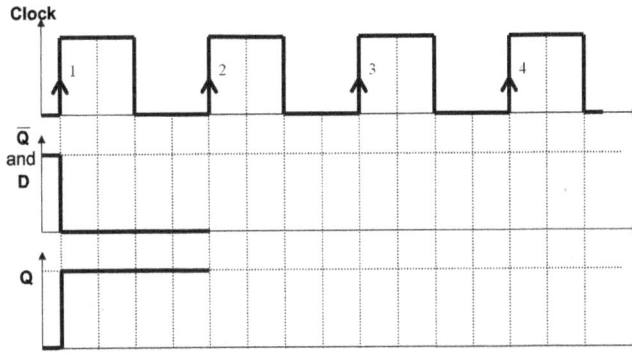

At the first rising edge (1), input D is at logic 1, since it is connected to $\overline{Q}$. This changes the Q output to logic 1, and so $\overline{Q}$ (and hence D) change to logic 0.

The outputs remain like this until the next rising edge of the clock (2).

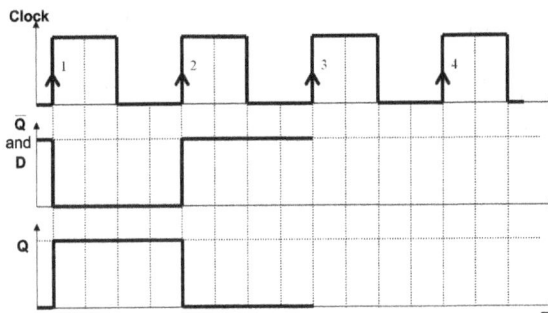

The D input is now at logic 0, which is copied to the output Q. This causes $\overline{Q}$ (and D) to change to logic 1 as shown below

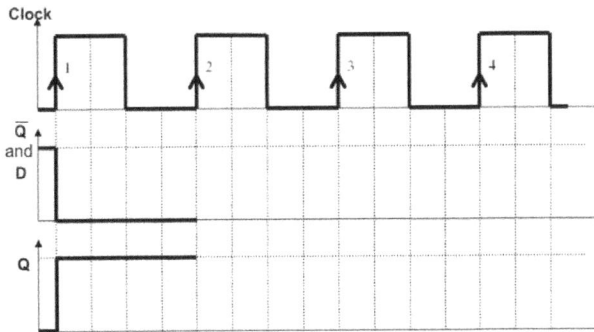

You should now start to see a pattern emerging in the diagram

At the next rising edge, the logic 1 present on the D input which is copied to the output Q, and so $\overline{Q}$ (and D) change to Logic 0 as shown below.

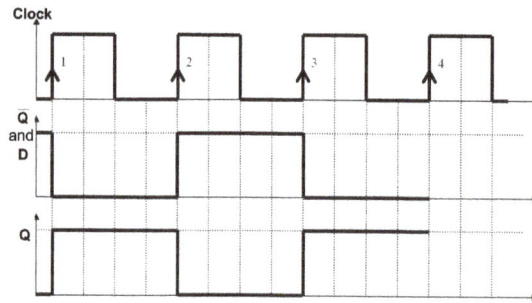

At the final rising edge, there is now logic 0 present on the D input. This is copied to the output Q, and $\overline{Q}$ (and D) change to logic 1 as shown below.

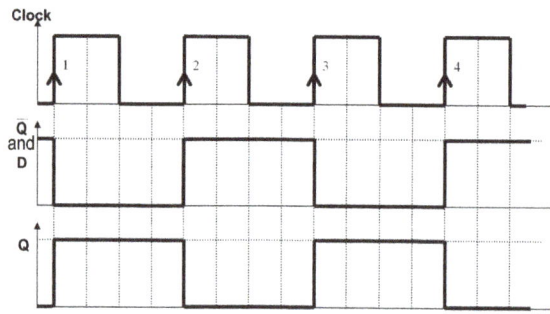

This pattern would continue in the same way for further clock cycles.

Notice that the output from Q has exactly half the frequency of the clock pulse, so this circuit could also be used as a simple frequency divider (the divide-by-two function).

## 2-bit Binary Up-counter

The divide-by-two action forms the basic building block of a binary counter. Another name for this circuit is a 1-bit counter.

Two 1-bit counters can be joined together as shown below to form a 2-bit binary up-counter:

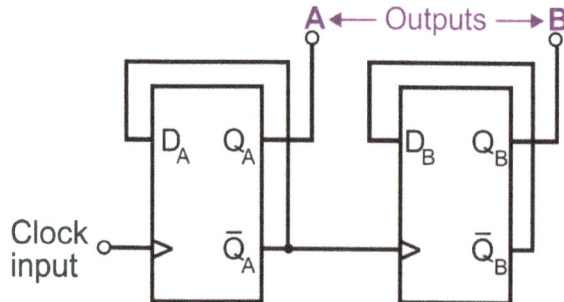

Notice that the clock input of the second counter is connected to the Q output of the first. The timing diagrams follow.

Initially QA and QB are at logic 0, Clock In = logic 0, $\overline{Q_A}$ = DA = logic 1, $\overline{Q_B}$ = DB = 1.

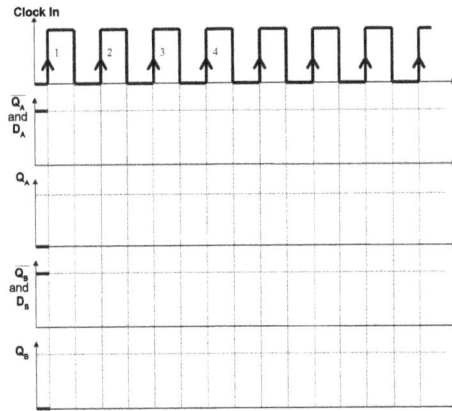

At the first rising edge of Clock:

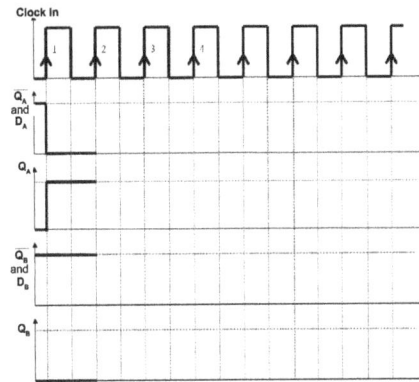

- DA (and $\overline{Q_A}$) = logic 1, so QA becomes logic 1

- so $\overline{Q_A}$ (and DA) becomes logic 0

- No change to QB, as the clock input to flip-flop B has gone from logic 1 to logic 0, which is a falling edge.

At the second rising edge of Clock:

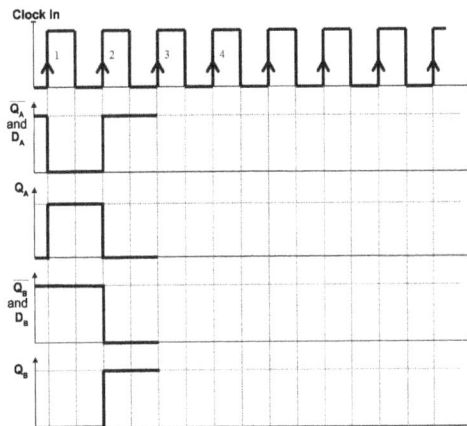

- QA becomes logic 0, since DA was logic 0

- $\overline{Q_A}$ is the opposite of QA and becomes logic 1

- QB becomes logic 1 as the clock input to flip-flop B has gone from logic 0 to logic 1, a rising edge, which copies DB to QB.

At the third rising edge of Clock:

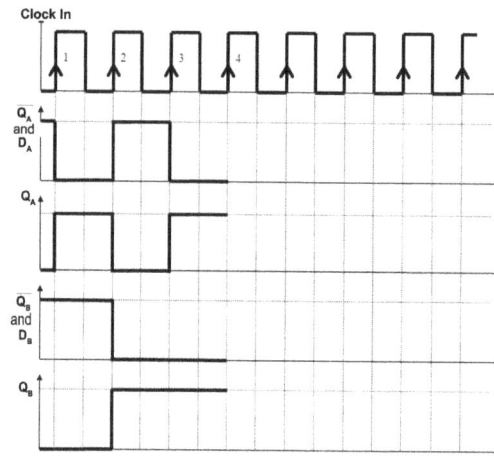

- QA becomes logic 1, since DA was logic 1

- so $\overline{Q_A}$ (and DA) become logic 0

- QB remains logic 1 as the clock input to flip-flop B went through a falling edge.

At the fourth rising edge of Clock:

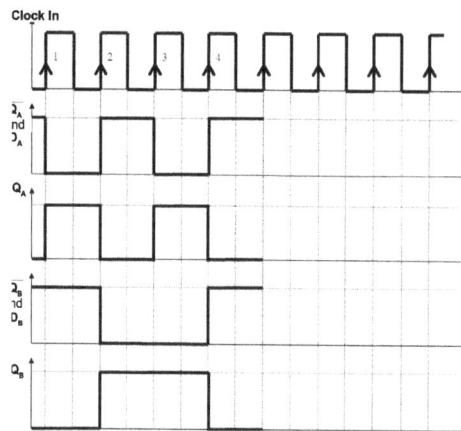

- QA becomes logic 0, since DA was logic 0

- $\overline{Q_A}$ is the opposite of QA and becomes logic 1

- QB copies DB and changes to logic 0, as the clock input to flip-flop B has gone from logic 0 to logic 1, a rising edge.

For further clock pulses, it is simply a case of repeating the patterns as follows:

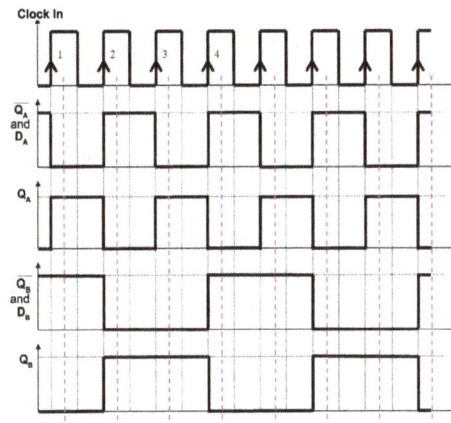

Looking at the values of QB and QA after each clock pulse shows us that we are counting up in binary.

The QA output is represented by the right-hand digit and is referred to as the least significant bit (LSB) as it is the one that changes the most often.

In the example at which we have just looked, there were a lot of graphs drawn but there was really no need to draw them all. Look at the final graphs.

- QB changes on the falling edge of QA (which is the rising edge of $\overline{Q_A}$)

- The output at QB is half the frequency of QA, or a quarter of the frequency of the clock, so the counter is a very good frequency divider.

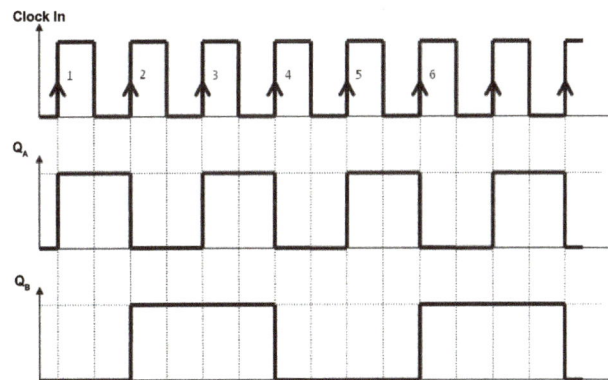

We have made a 2-bit counter, with A as the least significant bit, and B the most significant bit.

## Clocking

In the world of analog to digital and digital to analog conversion, the digital signal needs to be clocked accurately to prevent nasty sounding distortion. To take analog audio and make it digital, we simply take periodic samples of the audio signals amplitude and then send on as a series of ones and zeros which a computer or other digital device can use. Samples are taken many times a second, 44,100 times a second in the case of a CD (often referred to as 44.1kHz). These samples need to be taken at specific steady intervals so that when the data is converted back into analog data, we can be sure that it can be accurately recreated. Having a steady clock signal controlling your convertors is therefore crucial to ensure perfect capture of the audio and then nice clean playback afterwards. The accuracy of the clock is known as its 'Jitter' and is measured in nanoseconds or less. The smaller the jitter amount, the more accurately the capture and playback will be and the nicer the audio will sound. If the clock slowly changes over time then this is known as drift and just like Jitter, can cause your audio to not be captured or played back accurately and cause distortion.

## Clock key to Synchronous Systems

- Clocks help the design of FSM where outputs depend on both input and previous states.

- Clock signals provide reference points in time - define what is previous state, current state and next state:

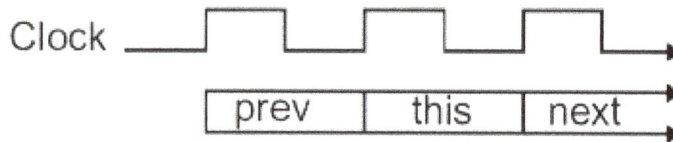

## Latch vs Flip-flop

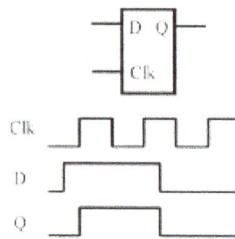

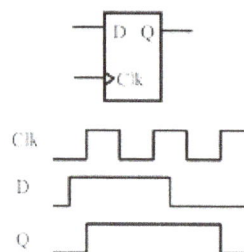

## Clock for Timing Synchronization

Clocks serve to slow down signals that are too fast.

1. Flip-flops/latches act as barriers.

2. With a latch, a signal can't propagate through until the clock is high.

3. With a Flip-flop, the signal only propagates through on the rising edge.

4. All real flip-flops consist of two latch like elements (master and slave latch).

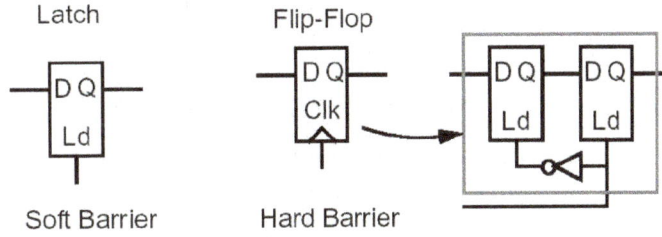

## Latch Timing Parameters

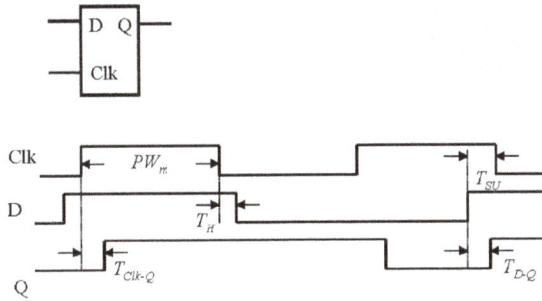

## Flip-flop Timing Parameters

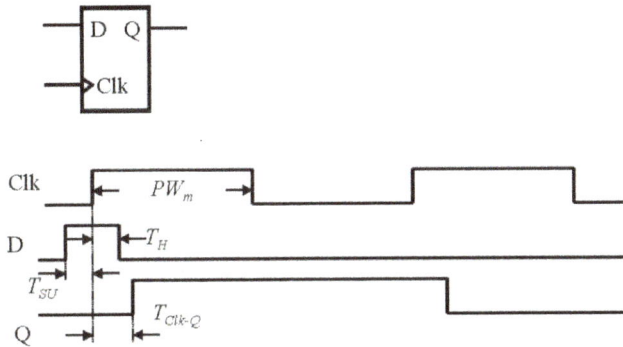

## Typical Clock System

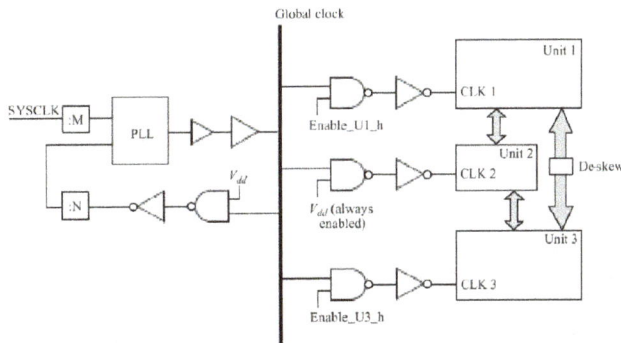

## Clocking Overhead

- Latches and flops slow down the slow signals
- Flip-flop delays the slowest signal by the setup + clk-q delay
- Latches delay the late arriving signals by the delay through the latch

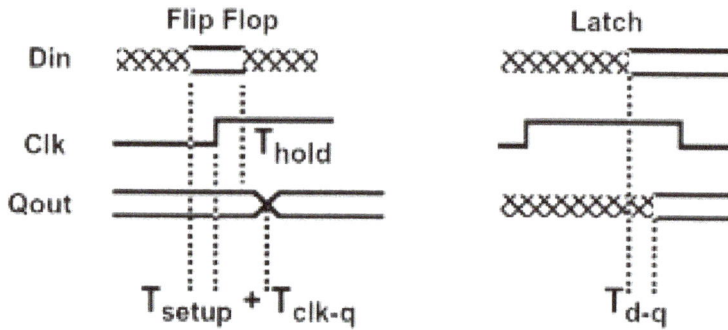

## Problem of Clock Skew

- Not all clocks arrive at the same time.
- Some clocks might be gated (ANDed with a control signal) or buffered There is an RC delay associated with clock wire.

Causes two problems
- ◆ The cycle time gets longer by the skew

$$T_{cycle} = T_d + T_{setup} + T_{clk-q} + T_{skew}$$

- ◆ The part can get the wrong answer

$$T_{skew} > T_{clk-q} - T_{hold}$$

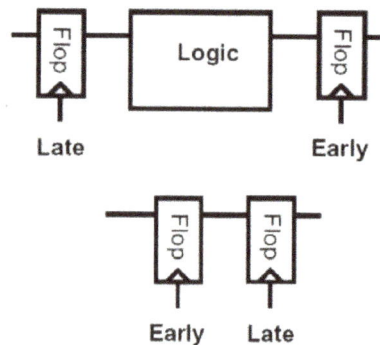

## Clock Skew and Jitter

## Clock Skew

- Spatial variation in temporally equivalent clock edges; deterministic + random, tSK.

## Clock Jitter

- Temporal variations in consecutive edges of the clock signal; modulation +random noise - Cycle-to-cycle (short-term) tJS Long term tJL.

Both skew and jitter affect the effective cycle time only skew affects the race margin

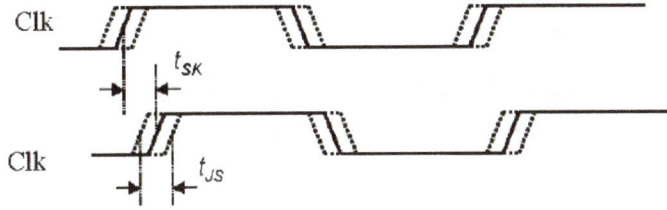

## Longest Logic Path-edge Triggered

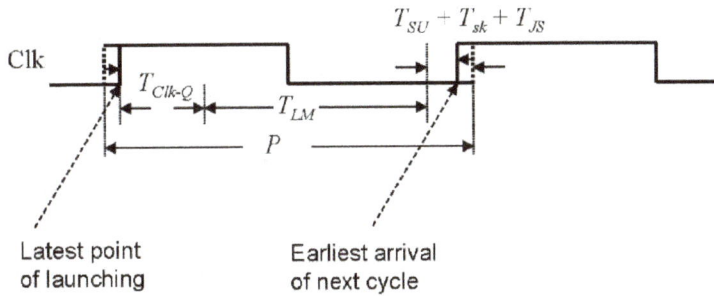

Latest point of launching

Earliest arrival of next cycle

$$P \geq T_{clk-QM} + T_{LM} + T_{SU} + T_{sk} + T_{JS}$$

## Shortest Path Constraint

If launching edge is early and receiving edge is late:

$$-T_{clk-Qm} + T_{Lm} \geq T_{sk} + T_H$$

$$T_{Lm} \geq T_{sk} + T_H - T_{clk-Qm}$$

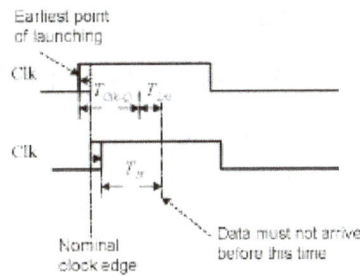

## Clocking Strategies

- Tradeoff between overhead/robustness/complexity

- Constraints on the logic vs. Constraints on the clocks

- Look at a number of different clocking methods:

  ◦ Pulse mode clocking.

  ◦ Edge triggered clocking.

- ◦   Two phase clocking.
- ◦   Single phase clocking.
- •   We will only look at system level strategy - consider clocked circuits in the next lecture.

## Pulse Mode Clocking

## Two Requirements:

1.   All loops of logic are broken by a single latch.

2.   The clock is a narrow pulse.

It must be shorter than the shortest path through the logic

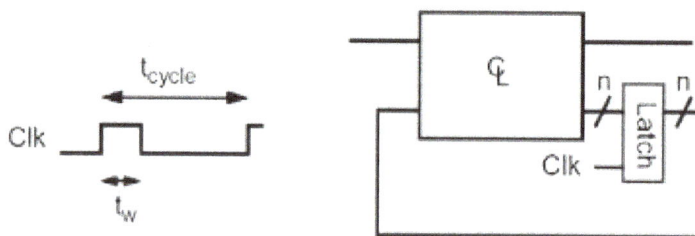

Timing Requirements

$$t_{dmax} < t_{cycle} - t_{d\text{-}q} - t_{skew}$$
$$t_{dmin} > t_w - t_{d\text{-}q} + t_{skew}$$

- •   Used in the original Cray computers (ECL machines)
- •   Advantage is it has a very small clocking overhead:

    One latch delay added to cycle

- •   Leads to double sided timing constraints

    If logic is too slow OR too fast, the system will fail

- •   Pulse width is critical

    Hard to maintain narrow pulses through inverter chains

- •   People are starting to use this type of clocking for MOS circuits

    - ◦   Pulse generation is done in each latch.
    - ◦   Clock distributed is 50% duty cycle
    - ◦   CAD tools check min delay

- •   Not a good clocking strategy for a beginning designer

## Edge Trigger Flip-flop

- Popular TTL design style.

- Used in many ASIC designs (Gate Arrays and Std Cells).

- Using a single clock, but replaces latches with flip-flops.

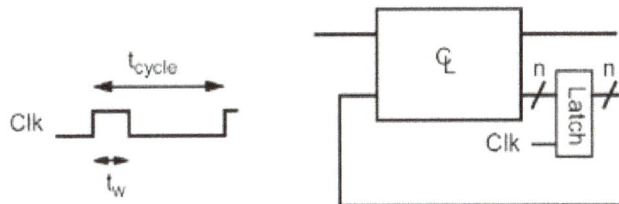

- Timing Constraints.

$$tdmax < tcycle - tsetup - tclk - q - tskew$$
$$tdmin > tskew + thold - tclk - q$$

- If skew is large enough, still have two sided timing constraints.

## Two Phase Clocking

Use different edges for latching the data and changing the output.

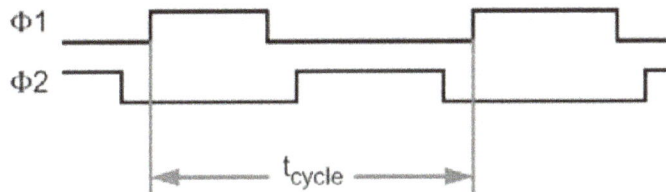

There are 4 different time periods, all under user control:

- $\Phi1$ high

- $\Phi1$ falling to $\Phi2$ rising

- $\Phi2$ high

- $\Phi2$ falling to $\Phi1$ rising

Look at shift register again:

1. If there is a large skew on the $\Phi2x$ clock, then the spacing between $\Phi1$ and $\Phi2$.

2. Can be increased to make sure that even with the skew, the $\Phi2$ latch closes.

3. Before the $\Phi1$ latch lets the new data pass.

4.  For some setting of the timing of the clock edges, the circuit will work.

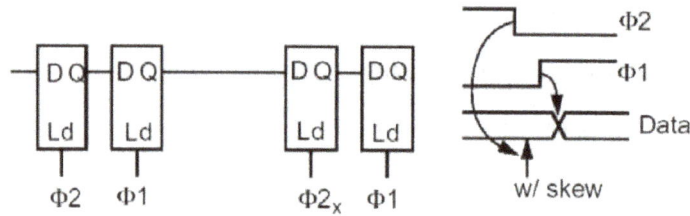

## Stable Signal Type

- We will give signals timing types, so it will be easier to know which latch to use:
- Output of a $\Phi 1$ latch is stable $\Phi 2$ (_s2) – good input to $\Phi 2$ latch
- Output of a $\Phi 2$ latch is stable $\Phi 1$ (_s1) – good input to $\Phi 1$ latch
- Signal is called stable2, since it is stable for the entire $\Phi 2$ period

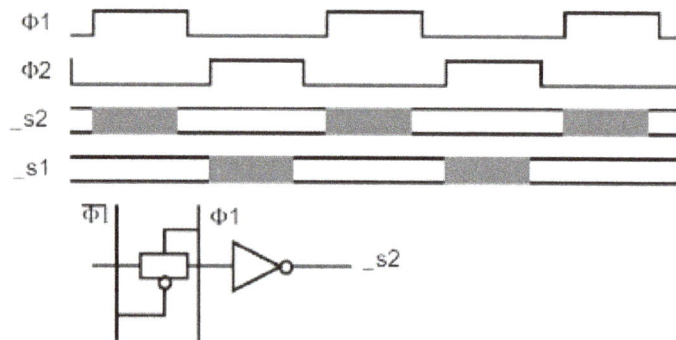

## General Two Phase System

- Combination logic does not change the value of timing types.
- No static feedback in the combination logic is allowed either. This makes the system not sensitive to logic glitches.

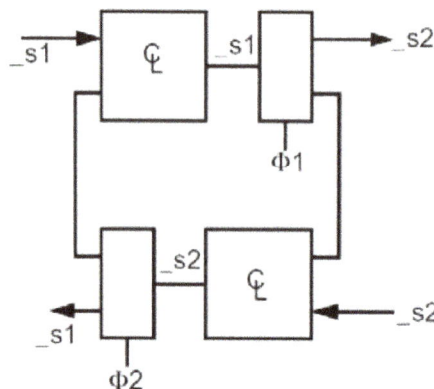

## Importance of Two Phase Clocking

## It is a Constrained Clocking Style

- Synchronous design

- Two clocks

- Constrained composition rules

But gives this guarantee:

## If you Clock it Slow Enough

- It will be a level sensitive design

- No race, glitch, or hazard problems

- No skew problems

- One sided timing constraints

- Impossible for logic to be too fast

## Mealey and Moore Machines

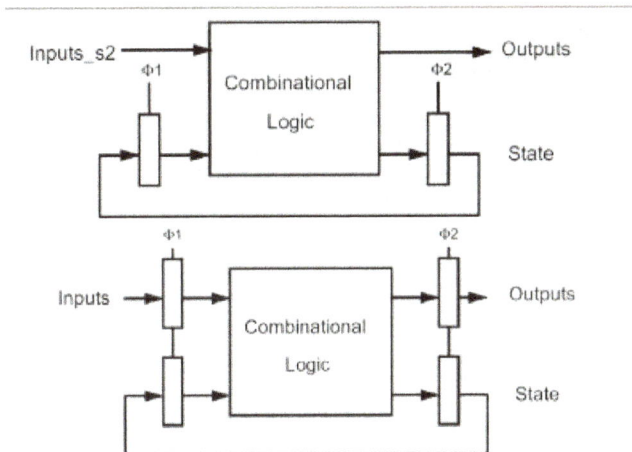

## More on Latch Timing

- Look a little more closely at latches, to come up with a more complete set of timing types (more than _s1 _s2 signals) that we can use in our synchronous designs.

- We should look at a latch since this the critical element.

- What is the weakest requirement on the input to a latch?

- Signal must settle before Φ1 falls, and not change for some time after Φ1 falling, even for a skewed Φ1 (this is usually called the setup and hold times of the latch).

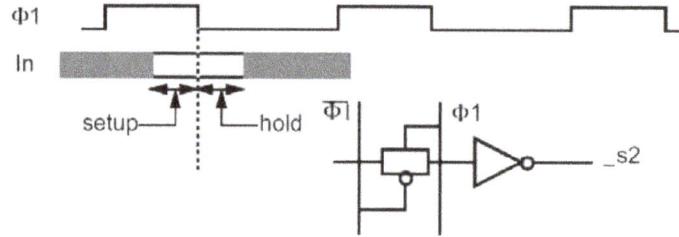

## Valid Signal Type

- The weakest input to a latch is called a valid signal (_v1 _v2).

- For a valid signal we need to be sure we can guarantee it meets the setup and hold requirements of the latch.

- To do this we need to have the signal settle off an edge that comes before Φ1 falling. The closest edge is Φ1 rising.

- The signal should not change until an edge occurs that comes after Φ1 falling. The closest edge is Φ2 rising.

- If we changed the input on Φ1 falling, most of the time the circuit would work fine. But if it failed, we can't change the clock timing to make the circuit work -- Φ1 falling controls the changing of the input, and the closing of the latch. Since we can't guarantee it would be ok a signal that changes on Φ1 falling would not be a _v1 signal.

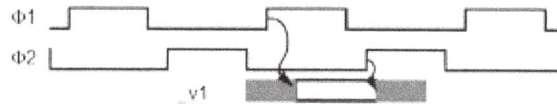

## Use of Valid Signal

- Very useful for pre charged logic.

- Is not needed for standard combinational logic with latches.

- This should always give stable signals.

- Can't use stable signals if you want to drive two signals/cycle on a wire (multiplex the wire), since the value has to change twice. There are many wrong ways to do it, and only one right way, which is shown below. The values become _v signals.

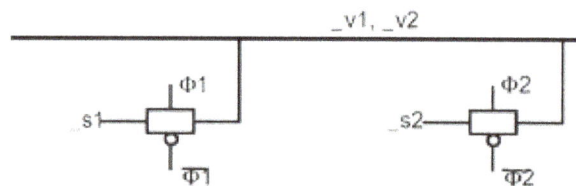

## Clock Gating

Clock signal is the highest frequency toggling signal in any SoC. the capacitive load power component of the dynamic power is directly proportional to the switching frequency of the devices. This implies that clock path cells would contribute maximum to the dynamic power consumption in the SoC.

Power consumption in the clock paths alone contribute to more than 50% of the total dynamic power consumed within modern SoCs. Power being a very critical aspect of the design, one need to make prudent efforts to reduce this. Clock Gating is one such method.

Clock feeds the clock pins all the Flip-Flops in the design. Clock Tree itself comprises of clock tree buffers which are needed to maintain a sharp slew (numerically small) in the clock path.

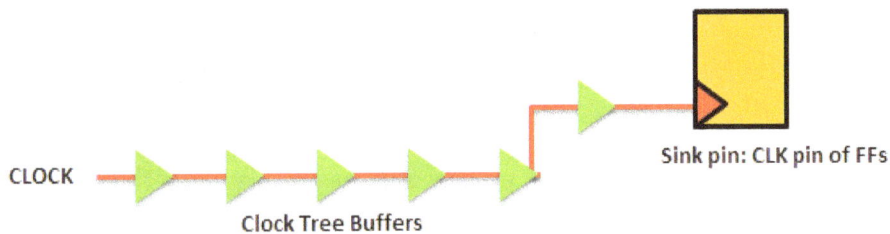

Consider the above figure. It is not necessary that the output of the flip-flop would be switching at all times. Modern devices support various low-power modes in which only a certain part of your SoC is working. This may include some key features pertaining to security or some critical functional aspects of your device. Apart from this, there are some configuration registers in your device which need to be programmed either once or very seldom. So, let's say, the above FF will not be switching states for a considerable period of time. If it is used the way it is, what's the problem? Power. Clock is switching incessantly. Clock Tree buffers are switching states and hence consuming power. So are the FFs. Remember that FF itself is made up of latches. So, despite the fact that input and output of the FF is not switching, some part of the latch is switching and consuming power.

What could be done to alleviate the above problem? Clock Gating is one such solution. Here's how it'll help.

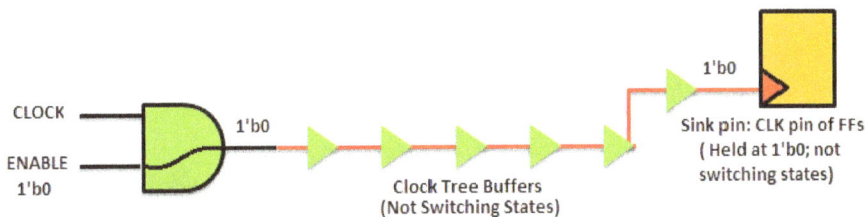

If you place an AND gate at the clock path and knowing that you don't need a certain part of your device to receive clock, drive a logic '0' on the ENABLE pin. This would ensure that all the Clock Tree buffers and the sink pin of the FF are held at a constant value (0 in this case). Hence these cells would not contribute to dynamic power dissipation. However, they would still consume leakage power.

Similarly, you can place an OR gate and drive it's one input to logic 1. Again, you would save on the dynamic power.

However, a word of caution, the output of the AND gate feeding the entire clock path might be glitchy. See the following figure:

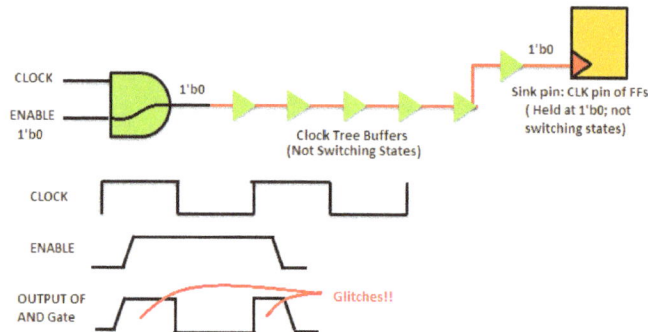

Solution: The output won't be glitchy if the enable signal changes only when the CLOCK signal is low. So, all you gotta make sure is that ENABLE is generated by a negative-edge triggered FF. This would ensure that the signal is changing after the fall edge of the CLOCK signal.

Similarly, while using an OR-gate, clock pulse would be propagated if the ENABLE signal changes when the CLOCK is high. Make sure that it is generated by a positive-edge triggered FF in order to avoid any glitch being passed onto the FFs.

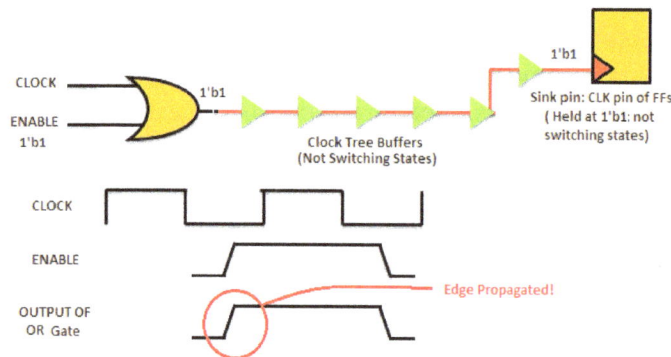

The reason a glitch would be detrimental anyway is because, Glitches constitute an edge. FF might sample the value because they are edge-triggered. But, problem is that all FFs have a certain duty cycle requirement (Also called Pulse-width check), which needs to be fulfilled in order to ensure that they don't go into METASTABILITY. And if an unknown state: X is propagated in a design, the entire functionality of the chip can go haywire.

- AND/NAND gate based clock gating is referred to as Active-High Clock Gating.

- OR/NOR gate based clock gating is referred to as Active-Low Clock Gating.

NAND and NOR clock gates work similar to AND and OR respectively.

So, Clock Gating is an efficient solution to save dynamic power consumption in the design. Modern SoCs have many IPs integrated together. Placing a clock gate and enabling them in various possible combinations is what gives rise to different low-power modes in the device.

# Clock Generator

A clock generator is a type of circuit that produces a continuous, synchronized electrical signal for timing purposes in a wide variety of devices. Because many high end electronic equipment require that electrical signals and mechanical devices work together in an efficient manner, clock generators are often a necessary component in many devices in order to ensure that all other components work harmoniously.

## Working of Clock Generators

Clock generators are generally made of a quartz or ceramic piezo-electric circuit board that includes an oscillator and an amplifier. As the piezo-electric material responds to changes in pressure, the oscillator produces a constant wave that repeats itself, such as a square wave, in order to synchronize external events. The amplifier receives and inverses this signal, passes it along to the output, and returns a portion of the signal back to the oscillator.

## Advantages

Clock generators are advantageous because they allow mechanical devices to stay synchronized with their digital counterparts. Many clock generators, known as "programmable clock generators," can be modified to change the signal they produce, allowing users to change the speed at which mechanical and digital devices perform tasks. Clock generators are usually small, lightweight, and inexpensive to produce, allowing them to be placed in ever smaller electronic devices such as laptops, notebooks, and smartphones.

## Applications

Clock generators can be used in a wide variety of applications, the most notable being computer systems. Clock generators are used in computers to manage memory cards, peripheral devices, CPUs, ports, etc. In fact, computer experts often reset clock generators in order to control these devices' speed and performance. Clock generators are also used in telecommunication systems, digital switching systems, and many mechanical devices.

## Timing-signal Generators (TSGs)

TSGs are clocks that are used throughout service-provider networks, frequently as the building integrated timing supply (BITS) for a central office.

Digital switching systems and some transmission systems (e.g., SONET) depend on reliable, high-quality synchronization (or timing) to prevent impairments. To provide this, most service providers utilize interoffice synchronization distribution networks based on the stratum hierarchy and implement the BITS concept to meet intraoffice synchronization needs.

A TSG is clock equipment that accepts input timing reference signals and generates output timing reference signals. The input reference signals can be either DS1 or composite-clock (CC) signals, and the output signals can also be DS1 or CC signals (or both). A TSG is made up of the six components listed below:

1.  An input timing interface that accepts DS1 or CC input signals.

2.  A timing-generation component that creates the timing signals used by the output timing-distribution component.

3.  An output timing distribution component that utilizes the timing signals from the timing-generation component to create multiple DS1 and CC output signals.

4.  A performance-monitoring (PM) component that monitors the timing characteristics of the input signals.

5.  An alarm interface that connects to the central-office (CO) alarm-monitoring system.

6.  An operations interface for local craftsperson use and communications with remote operations systems.

## Clock Signal

A digital clock signal is basically a square wave voltage similar as the one shown below:

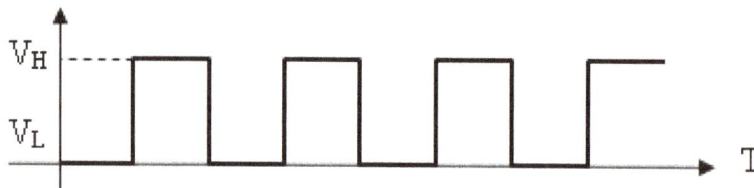

As shown, it has only two levels, one is zero and the other one is high, which the high level can be different according to the requirement of the circuit. For example the high level in TTL standard is 5V. This signal mostly has a 50% duty cycle, meaning that the duration of a high and a zero is the same. The frequency of the clock can be anything as needed by the digital circuit that is using it. But what is a clock used for?

Digital circuits always have some input and generate digital outputs accordingly. Some digital circuits are not clocked, meaning that the input applied to the circuit flows through digital gates without any timing or storage and generates the output. It only takes a time equal to the propagation delay time to reach the output.

On the other hand most of the digital circuits that do more complex processing on the digital inputs such as controllers, processors or state machines are timed and the signal can't just go through. In these circuits a clock with a fixed frequency is used for timing. A clock plays very important role as it is used to open and close digital paths, allow or stop a process and in general provide timing for the circuit. You can compare a clock with the traffic lights. They stop and allow the traffic in a timely manner so that the traffic can flow smoothly with the least delays. If you just let the traffic through there will be a big jam and the output is unpredictable.

Clocks are especially used for digital circuits with feedbacks and also to avoid glitches in a circuit. What is a glitch? A glitch is an unpredictable output. Say you have some input and for those you expect a known output, but before the output settles to what you expect, you might have one or more transitions that are not suppose to be there. These are called glitches. They happen because the inputs have to go through different gates and the propagation delay of each can be different. Therefore the results arrive to the final gates in different times. This difference in data arrival results in changes in the output until all signals settle and the output is valid. If glitches are not eliminated, they will go to the next stage of the circuit and generate more unpredictable results. To avoid them a clock can be used to time the signals. Assume the inputs to the circuit are provided with one rising edge of the clock and the output of the circuit is read by the next rising edge of the clock. If the period of the clock signal is higher than the total propagation delay of the circuit, the output will be read when it is completely settled and therefore no glitch happens.

To store and pass the data or digital signals through, some specific gates are used which are called latches or flip-flops. These are some kind of memory that store their input over their output by a specific level or edge of the clock.

A clock with a higher speed will allow a faster process and that's why we see the increase in the clock frequency in computers and processors every day. But you may ask why won't we just increase it to the maximum already? Well it is not that easy. As mentioned before clock has to wait enough time for all the data to be ready and then pass them forward. This means that the digital circuit delay becomes a very important factor in all this. If the circuit delay is high, the clock has to be slow to allow enough time for the data to arrive. Therefore increasing the clock frequency is only possible if the circuit delay decreases and that is a very hard thing to do in most cases.

## Clock Generation

A clock generator is used to generate a clock, which is an oscillator that provides a square wave output. An oscillator circuit always has a feedback that makes the circuit oscillate. This feedback also provides parameters that can make a certain frequency. There are many different ways to make an oscillator. Below you can see two of the famous ones. First is to use a simple Inverter plus a feedback component which usually is a crystal.

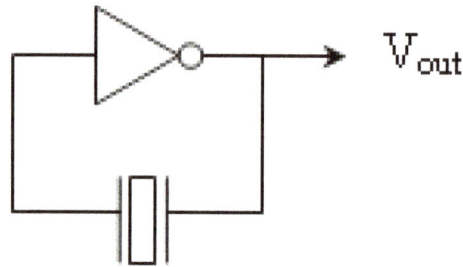

Crystal is an electrical component that is a very precise filter allowing only a very specific frequency through. It usually looks something like below, well it can come in any different size and form.

They call it crystal because inside the package there is a piezoelectric crystal with a very specific size and weight. A piezoelectric crystal, when voltage is applied to it, can resonate with a very precise frequency that depends on its size and properties and allows only signals with its resonance frequency though. That's why it is used to filter any other frequency and provide a fixes frequency signal. The resonance frequency of the crystal is written on its package.

In the Circuit shown above, first assume the input of the inverter goes low and therefore the output goes high. This transition which is usually very sharp contains many different frequency harmonics, but only the one with the same frequency as the crystal passes through. So the high goes to the inverter input and its output drops to low. This happens continuously and therefore resulting in a square wave with a fixed frequency.

Below is another circuit using a Schmitt_Triggered inverter.

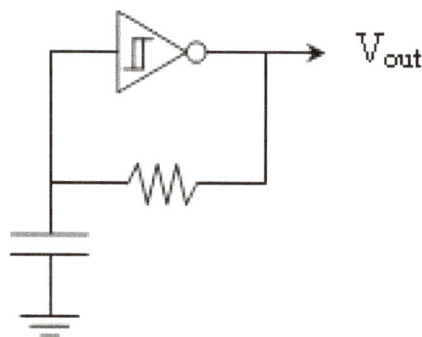

A Schmitt-Triggered inverter compares the input signal with two thresholds instead of one as in simple inverters or gates, the input has to be higher than one threshold for the output to change, and lower than the other one for the output to change back. It generates a hysteresis pattern which is showed in the figure below for the inverter:

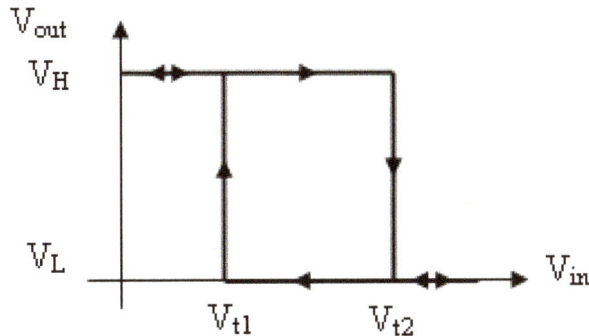

Arrows in the graph show the direction of the input voltage change. When there is a low on the input of the inverter and a high at its output, the capacitor on the input of the inverter is charged through the feedback resistor. When the capacitor is charged high enough that the input voltage of the inverter passes the higher threshold, then it's output drops to zero. Now the capacitor starts discharging through the same resistor and when its voltage drops below the lower threshold of the inverter, its output jumps up again and this action keeps happening resulting in the clock signal. Figure below shows the input and output waveforms of the inverter.

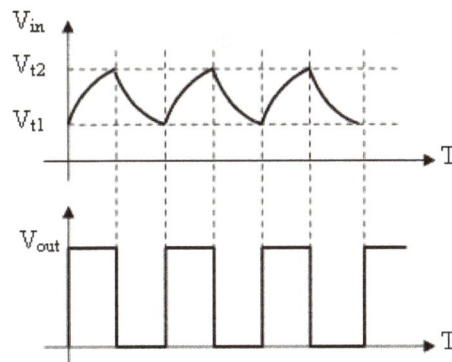

The oscillating frequency of this circuit depends on how fast the capacitor charges and discharges, and therefore is a factor of the resistance and capacitance values, while it is also dependent on the threshold levels.

## Power Optimization

Reduction in Power dissipation is an essential design issue in VLSI circuit. The design parameters have conflicting affect on overall performance of the system. Depending upon the component and function, different optimization approaches can be adopted. For instance power consumption in multiplier is data dependent as gate switching activity contribute to more power consumption. The gate switching activity can be optimized by considering different gate combinations. Gate switching activity can be reduced by employing various algorithms. For instance, in multiplier, design method used for multiplication have affect on power consumption. In multiplier, besides primary school method of bit multiplication, Booth algorithm & Modified Booth algorithm can be used for efficient multiplication.

## Dynamics of Power Consumption

Power dissipation in VLSI circuits is due to three major sources i.e., power required to charge or discharge a node, power dissipation due to output transition and power dissipation due to leakage current. So optimizations can be achieved by concentrating any one or combination of above design issues. The power consumption can be given by following equation:

$$P_g = \hat{f} \, C_{SC} \, V_{dd}^2$$

Where,

$P_g$ = Power consumed by a single gate

$\hat{f}$ = Average operating frequency of the gate

$C_{SC}$ = Switching capacitance of the gate

$V_{dd}^2$ = Power supply voltage

Using (Equation above) we obtained; for any number of gates in the chip.

$$P_g = N \, \hat{f} \, C_{SC} \, V_{dd}^2$$

Where,

$N \, \hat{f}$, represent the total number of bit operations per second.

## Power Dissipation Estimation

In all logic circuits, power consumption is related to information transfer and each circuit have inherent requirement of information transfer. Let R is the transfer rate requirement for a given architecture; the lower bound of this rate may be determined which consequently can be used for power dissipation estimation. The different digital architecture can perform same function but may have different information transfer rate and different channel capacity. Channel capacity can be given as:

$$C = \int_0^W \log_2 \left[ 1 + CNR(f) \right] df$$

Where, SNR gives signal to noise ratio. For any meaningful transfer capacity should be greater or equal to R. The overall noise power in digital circuit is a function of signal power, temperature, semiconductor property etc. However, power dissipation is mainly due to ground bounce. The lower bound of the power dissipation can be calculated using information transfer capacity of channel. Let R is required information transfer, W is the channel bandwidth, sigma be noise power and C is channel capacity. Using (Equation above), lower bound of power dissipation can be given as:

$$p_{D1\min,ser} = C_L\left(2^{\frac{R}{W}} - 1\right)8\sigma_N^2 W$$

## Data Dependent Power Optimization

Complexity of data contributes to gate switching activity in the circuit. By adopting efficient computation algorithm design components of the circuit at gate level can be reduced. Investigating different designs & arithmetic representations might reduce power variations. Model can be built to simulate power consumption.

By applying simulation to standard designs and comparing with optimum, better design component can be found. Gate switching for all initial states and all inputs can be simulated to analyze power consumption in each option. Data dependence consideration is helpful in gate design complexity. Ordering of gate inputs affect both power and delay. Parsed has described methods to optimize the power and delay of logic gates based on transistor reordering. Therefore, considerable improvements in power and delay can be obtained by proper ordering of transistors. For instance, late arriving signals can be placed closer to the output to minimize gate propagation delay.

Another approach to reduce power is to consider the size of gate, which has significant impact on circuit delay and power dissipation. By increasing the size of transistors in a given gate, delay of the gate can be decreased but in contrast, power dissipated in the gate and fabrication space increases. Therefore, an optimum balance can be achieved by sizing of transistors appropriately. A method is to compute the slack at each gate in the circuit, where the slack of a gate corresponds to how much the gate can be slowed down without affecting the critical delay of the circuit. Alternatively, in different sub – circuits, where slack is greater than zero are utilized and the size of the transistors is reduced until the slack becomes zero, or the transistors attain a minimum size.

## Combinational Gate Level Design

In gate level design of circuit, different combination of logical gates may produce same circuit output but different value of power consumption. Path balancing, factorization and don't care optimization may be utilized to optimize power consumption. Path balancing can be achieved by avoiding delay at each input gate.

Genetic Algorithm can be used to determine different combination of gates and power consumption can be formulated by devising Fitness Function. Coello has proposed design of combinational logic circuit based on Genetic Algorithm. By defining chromosome development scheme of various combinations of logic circuits can be evolved using cross over and mutation. This approach is more efficient (in some particular scenarios & constraints) than human designer as various constraints of design circuit can be devised subject to fitness function.

Genetic Algorithm can reduce number of gates, which consequently reduce power consumption; as the work of Coello shows on 2-bit adder and 2-bit multiplier with a particular 'cardinality'; about 56% reduction in number of gates for the circuit can be achieved.

Number of gates versus power saving in cmos – based on iscas-89 benchmark circuits

It is very obvious that number of gates is directly proportional to the power consumed; as shown in the (Graph). So, if we minimize the number of gates in the design we can achieve low power consumption as well; as Coello proposed.

It is very obvious that number of

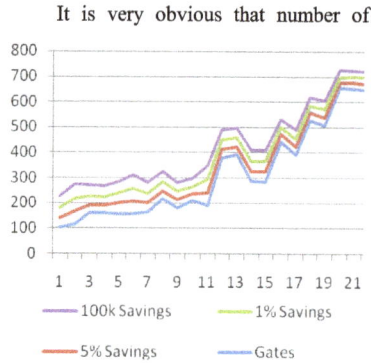

Graph: Gate Vs Power based on the work.

On the other hand power reduction is sometimes a "Give n Take" scenario; if we achieve reduction in power supplied there may be some loss in speed, efficiency etc., and so, minimizing power without losing other recourse parameters is the need of time.

| Power Reduction in supply voltage | Speed Loss | Constraints/ Specifications |
|---|---|---|
| leakage power reductions up to 54% | Not reported | "Logic design to reduce the leakage power of CMOS circuits that use clock gating to reduce the dynamic power dissipation tested on ISCAS-89 benchmark circuits" |
| 0.13 V or 800 times | 19 times | "0.5-fim gate length and static logic" |
| "1.1 V supply and consumes less than 5 mW-which is more than three orders of magnitude lower power compared to equivalent commercial solutions." | Not reported | --- |

# References

- Sequential-machine, dictionaries-thesauruses-pictures-and-press-releases: encyclopedia.com, Retrieved 23 June 2018

- Sequential-circuits-basics: electronicshub.org, Retrieved 13 July 2018

- Digital-electronics-flip-flop-circuit-types-and-applications: elprocus.com, Retrieved 29 May 2018

- Flip-flops-types-applications-woking, digital-electronics: daenotes.com, Retrieved 19 April 2018

- What-is-clocking: audient.com, Retrieved 11 March 2018

- Clocking-strategies-in-vlsi-systems: vetechlib.blogspot.com, Retrieved 21 June 2018

# 4

# Floorplanning

The floorplan of an integrated circuit is a schematic representation of the arrangement of functional elements. This chapter includes vital topics related to wire bonding, line driver and floor plan, which are crucial for an extensive understanding of floorplanning.

## Floorplan

A floorplanning is the process of placing blocks/macros in the chip/core area, thereby determining the routing areas between them. Floorplan determines the size of die and creates wire tracks for placement of standard cells. It creates power ground (PG) connections. It also determines the I/O pin/pad placement information.

A typical floorplanning includes:

1. Define the size of your chip/block and Aspect ratio

2. Defining the core area and IO core spacing

3. Defining ports specified by top level engineer.

4. Design a Floor Plan and Power Network with horizontal metal layer such that the total IR Drop must be less than 5% (VDD+VSS) of VDD to operate within the power budget.

5. IO Placement/Pin placement

6. Allocates power routing resources

7. Place the hard macros (fly-line analysis) and reserve space for standard cells. (Please refer rules for placing hard macros)

8. Defining Placement and Routing blockages blockages

9. If we have multi height cells in the reference library separate placement rows have to be provided for two different unit tiles.

10. Creating I/O Rings

11. Creating the Pad Ring for the Chip

12. Creating I/O Pin Rings for Blocks

Every subsequent stage like placement, routing and timing closure is dependent on how good your floorplan is. In a real time design, you go through many iterations before you arrive at an optimum floorplan.

Floorplanning takes in some of the geometrical constraints in a design. Examples of this are:

- Bonding pads for off-chip connections (often using wire bonding) are normally located at the circumference of the chip.

- Line drivers often have to be located as close to bonding pads as possible.

- Chip area is therefore in some cases given a minimum area in order to fit in the required number of pads.

- Areas are clustered in order to limit data paths thus frequently featuring defined structures such as cache RAM, multiplier, barrel shifter, line driver and arithmetic logic unit.

- Purchased intellectual property blocks (IP-blocks), such as a processor core, come in pre-defined area blocks.

- Some IP-blocks come with legal limitations such as permitting no routing of signals directly above the block.

Inputs for Floor Planning Stage:

- Synthesized Netlist (vhdl)

- Logical and Physical Libraries

- TLU+ Files

- Physical partitioning information of the design

- Design Constrains (SDC)

- Physical information of your design (rules for targeted technology)

- Floorplan parameters like height, width, utilization, aspect ratio etc.

- Pin/pad Position

Outputs of Floor Planning Stage:

- Die/Block area

- I/O pad/placed

- Macro placed

- Power grid design

- Power pre-routing

- Standard cell placement areas.

## Purpose of Floor Planning

The first step in the Physical Design flow is Floor Planning. Floorplanning is the process of identifying structures that should be placed close together, and allocating space for them in such a manner as to meet the sometimes conflicting goals of available space (cost of the chip), required performance, and the desire to have everything close to everything else.

Based on the area of the design and the hierarchy, a suitable floorplan is decided upon. Floor Planning takes into account the macro's used in the design, memory, other IP cores and their placement needs, the routing possibilities and also the area of the entire design. Floor planning also decides the IO structure, aspect ratio of the design. A bad floor-plan will lead to waste-age of die area and routing congestion.

In many design methodologies, Area and Speed are considered to be things that should be traded off against each other. The reason this is so is probably because there are limited routing resources, and the more routing resources that are used, the slower the design will operate. Optimizing for minimum area allows the design to use fewer resources, but also allows the sections of the design to be closer together. This leads to shorter interconnect distances, less routing resources to be used, faster end-to-end signal paths, and even faster and more consistent place and route times. Done correctly, there are no negatives to Floor-planning.

As a general rule, data-path sections benefit most from Floorplanning, and random logic, state machines and other non-structured logic can safely be left to the placer section of the place and route software.

Data paths are typically the areas of your design where multiple bits are processed in parallel with each bit being modified the same way with maybe some influence from adjacent bits. Example structures that make up data paths are Adders, Subtractors, Counters, Registers, and Muxes.

## Floor Planning Considerations

Gate count estimation:

> Rough estimation is needed to calculate die size.

Critical Port placement:

> In Top-Down approach port location is decided by considering all blocks location and communication between blocks.

> Moving ports a little does not vary but more leads to routing and timing issues.

## Logical Cell Types

Macro Placement:

i) Hard blocks or Memories are placed along edge of the block.

ii) Fly line analysis is needed when selecting macros because a design may contain more number of hard macros.

iii) Why space between block edge and macro edge?

If macros have pins on all sides then min spacing is required to provide sufficient routing channels to connect standard cells.

Memory Aspect Ratio:

Proper aspect ratio should provide by vendor otherwise core boundary blocks the pins and logic which talking to macro spreads all along which creates problem for setup timing.

## Memory Placement and Alignment

## Floor Planning Guide Lines

- Checking of net connection from macro to macro and macro to standard cells i.e., checking fly lines

- More number of connection from macro to macro place them near to each other most preferred is near core boundaries

- If i/p pin is connected to macro place near to pin or pad which is better

- More connections of macro to standard cells spread macro inside core area

- Avoid crisscross placement of macros in order to save routing resources as well as from routing ,Placement and congestion issues

- Soft Blockage: During placement that block is maintained free from standard cells and during optimization buffers are placed in that block in-order to meet timing.

- Hard Blockage: It doesn't allow any placement or during optimization.

- Partial Blockage: If it is given 20% .Then 20% of placement is blocked during placement Spacing is required between macros to avoid congestion around macros

- Go for iterative process of different floor plans when timing is not closing.

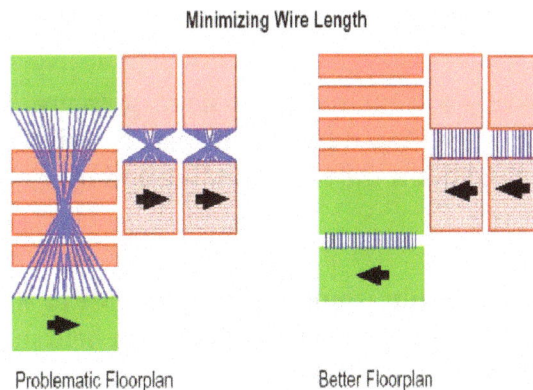

Minimizing Wire Length

Problematic Floorplan                    Better Floorplan

- Typically, macros are placed around edges of blocks, keeping one large main are for standard cells

- Leave a 'halo' of space between macros on all sides,

- For non-pin sides of macros a minimal separation is adequate,

- For pin sides of macros a larger separation is appropriate,

- Allow channels for routing, pin access and possible buffer insertion,

- Leave space between macros and edge of block, to allow for buffer insertion and power stripes to feed standard cell rows between macro and block edge.

# Wire Bonding

Since electronic components are becoming smaller and smaller and microelectronics is becoming the norm, the demand for wire bonding is rapidly increasing. The wire bonding process, which can be used in all types of integrated circuits and semiconductors, fixes the wires for necessary components by using a specific combination of heat, pressure, and energy.

The wire bonding method eliminates the need for high levels of heat that would be found when using other attachment methods, such as soldering. It is typically used with soft types of fine diameter wires, which are known as bondwires.

## Bondwires

Bondwires may be made of gold, silver, copper, or aluminum. Carefully selected alloys are also used in bondwires for their specific properties. These extremely fine wires can be as small as 15 micrometers and can be as large as hundreds of micrometers (when there are applications requiring higher system power).

Bondwires are usually made out of copper—but gold, aluminum, and silver are still popular options. These specific types of bondwire are carefully selected based on the design, requirements, and application of the component that contains them. The bondwire's material also has an impact on the cost and the production of the component in question.

## Types of Bonding

As mentioned, there is typically a combination of heat, pressure, and ultrasonic energy used in the wire bonding process. Therefore, different metals, as well as different specialty types of applications, will be better suited to the different bonding options.

The different bonding options are:

- *Thermosonic bonding*: uses ultrasonic energy and force to slightly heat the material and the wire, while pressing them together. Both are then held in place and vibrated for a set amount of time to solidify the connection.

- *Thermo compression bonding:* in this process, the contact surface is heated (and sometimes so is the wire). The two are then pressed together to create the bond. Ultrasonic energy and friction are not required in this method and are usually only used with gold wire on a gold surface.

- *Ultrasonic bonding*; also uses force and ultrasonic, but it does not use heat. The wire and the material are both at room temperature during the entire process. It can be used with any type of bondwire and similar metal with very good results.

Within each of the general bonding types, there are also specific bonding techniques. These techniques vary based on the application, the metal, and the type of wire used in the component.

## Leading Wire Bonding Applications

The most advanced wire bonding applications include: ultra-fine pitch (<60-µm pitch), stacked die, and multi-tier applications. These advanced applications often require more process optimization, as well as higher requirements for bonding material and equipment.

*Fine-pitch Applications:* Wire bonding fine-pitch capability has been demonstrated in the laboratory at 35-µm inline pitch. For 35-µm pitch ball bonding, 15-µm wire typically is used with a bonded ball diameter of 27 µm.

Fine-pitch applications demand a higher capability of the wire bonder, including better control of the bonding force, ultrasonic energy level, as well as a looping capability of fine wires, which has much less strength and is more inclined to loop sway. A wire bonder that meets the demands of fine pitch should also include high-precision motion and a vision system with submicron accuracy.

*Stacked Die Applications:* Stacked die applications are one of the fastest-growing trends in the semiconductor industry. The desire for smaller, lighter, and smarter devices is driving this 3-D packaging technology. Stacked die applications present various wire bonding challenges, including low-loop and multi-level wire bonding loop clearance requirements, bonding to overhang unsupported die edges, and loop resistance to wire sweep during molding.

Most wire bonding applications use the typical forward bonding process, because it is faster and more capable of finer pitch than reverse bonding. However, forward ball bonding has a loop height constraint due to the neck area above the ball. Excessive bending above the ball can cause neck cracks, which results in reliability problems. Reverse bonding can achieve loop height lower than 75 µm (figure below).

Figure: Reverse ball bonding loops.

*Multi-tiered ICs:* Traditional wire-bonded ICs have in-line bond pads on the periphery of the die. In the past, staggered pad designs (2-tier) were commonly used when I/O requirements exceeded silicon real estate. Increasing the number of pad rows (or tiers) is a natural progression in increasing the number of I/Os. Often found with high-end graphic and chip-set applications, tri-tier devices also are becoming more common. Early qualification of quad-tier devices are also beginning.

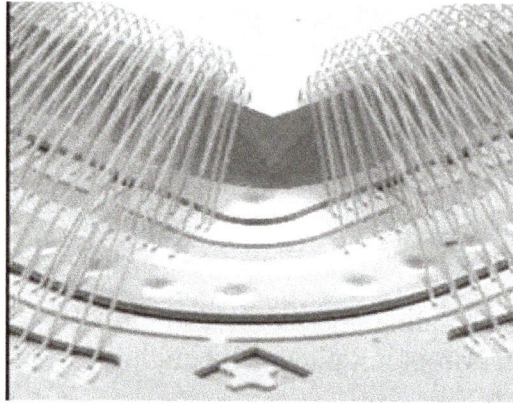

Figure: Multi-tier application.

Looping is the obvious challenge for the wire bonding process with multi-tiered IC applications. While in-line applications have single layers of loops over the first bond, multi-tiered devices require multiple layers of wire loops – ultimately driving up the overall package height. Loop heights can be as low as 50 to 75 μm for the shortest wires to more than 400 μm on the longest signal wires. In production today, gaps of 2 to 4 times the wire diameter are typical (with larger gaps in the upper layers) to ensure acceptable process yields across a population of wire bonders.

A second challenge for multi-tiered devices is associated with the first bond process. Adding bond pads without increasing the silicon area requires placement of active circuits below the bond pads. This, along with inclusion of low-$k$ interlayer dielectrics, will continue to challenge the first bond process. To bond to this type of package, the wire bonder needs fine control of impact and steady-state bond forces, along with careful control over USG applications.

## Wire Bonding Techniques

## Ball Bonding

Gold Ball Wire Bonding: Thermosonic tailless ball and stitch bonding is the most widely used assembly technique in the semiconductors to interconnect the internal circuitry of the die in the external world. This method is commonly called, Wire Bonding. It uses force, power, time, temperature, and ultrasonic energy (sometimes referred to as bonding parameters) to form both the ball and stitch bonds. Typically for the ball bond, the metallurgical interface is between gold (Au), and aluminum (Al) bond pad (typically with 1% silicon (Si) and 0.5% copper (Cu). As for the stitch bond, it is bonded to a copper alloy with thin silver (Ag) plating for lead frame. For BGA substrate, the stitch bond is bonded to copper/nickel base material with gold (Au) plating.

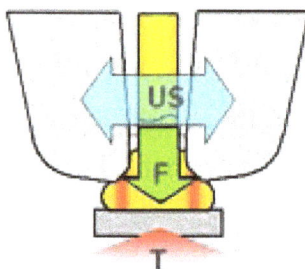

Ball bond formation                                            Stitch bond formation

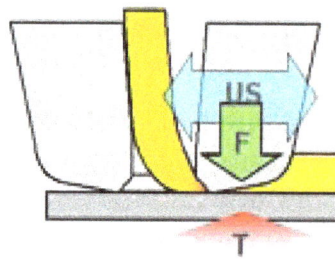

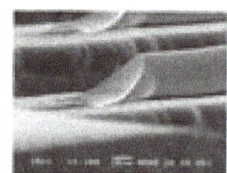

The ultrasonic transducer (typically for new generation of wire bonders, the piezoelectric element is >100KHz), which converts the electrical energy into mechanical energy, transmits this resonant energy at the tip of the bonding capillary. The capillary that is clamped perpendicularly to the axis of the transducer-tapered horn is usually driven in a y-axis direction vibration mode. Bonding capillaries are made of high-density Alumina ceramic material, $Al_2O_3$, typically 1/16" (.0625"/1.587mm) in diameter and .437" (11.10mm) in length. The final capillary design depends upon the package/device application and wire diameter to be used. To determine the correct capillary design in general, bond pad pitch (BPP), bond pad opening (BPO), target mashed ball diameter (MBD) are the essentials.

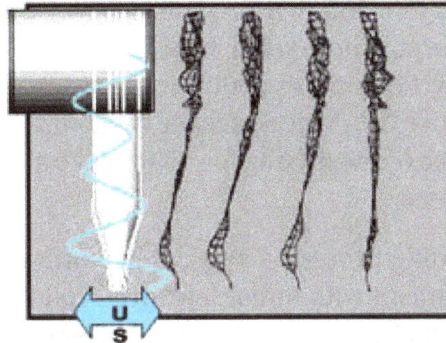

Variation characteristic of the capillary

A fine gold wire made of soft, face-centered-cubic metal (FCC), usually ranging from 18μm to 33μm in diameter (depending upon the device/ package application) is fed down through the capillary. It is usually characterized by its elongation (shear strain), and tensile strength (breaking load). Selection of the appropriate wire type to be used for a given application would be dependent on the specification of these elongation, and tensile strength. In general, the higher elongation (or higher strain), it means that the wire is more ductile. This is a good choice for low-loop, and short wire type of wire bonding application. If the requirement is for higher pull strength readings, a harder wire type having a higher tensile strength has to be considered. The small incursions of ultrasonic energy at the tip of the capillary are transmitted to the Au ball and down to the Al bond pad to form the ball bond. After which, the capillary lifts up and form the looping profile, and then comes down to form the stitch bond. This cycle is repeated until the next unit is bonded.

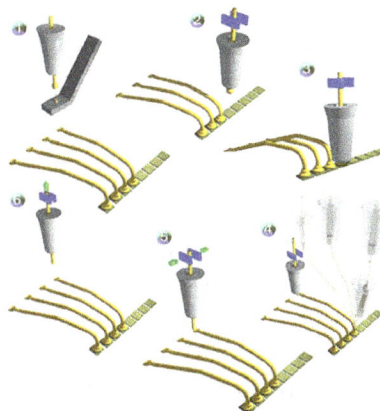

The bonding cycle

An intermetallic compound, Au-Al, is formed when the Au is bonded thermosonically to the Al bond pad metallization. The metallurgical interface of void free Au-Al formation has a significant increase in the shear strength readings of the ball bonds tested- provided that there are no impurities present in the bond interface even if it has been exposed to high temperatures. However, if the impurities are in the interface are welded poorly, the ball shear strength produces a significant degradation in its readings.

## Copper Wire Bonding

Wire bonding process is commonly used to interconnect chips to the outside world using gold wire since its inception in the mid 1950 using thermo-compression, an application of heat and force. However, it was not enough to form a more reliable oxide free ball and stitch bonds interface until the introduction of thermosonic bonding in 1960 incorporates ultrasonic energy. For decades, continuous progression to improve the device-package reliability has been the primary goal while cost of ownership has become one of the driving forces to make all the electronic gadgets available nowadays, affordable to the masses.

In general, the copper wire bonding process is very similar with gold wire bonding as it basically uses the same wire bonder equipment with minor hardware and software retrofits. Instead of gold wire, it is replaced by copper wire, though not limited; the range is typically from 15μm to 50μm in wire diameters depending upon the package-device application. Copper wire bonding offers significant advantages over gold – superior product performance in terms of electrical and thermal conductivity; better product reliability due to slower intermetallic growth that causes voids; and higher break load during wire pull testing.

One of the early day drawbacks of using copper wire in the wire bonding process is oxidation problem which can impact the reliability and integrity of the encapsulated device inside the electronic package. As we all know, oxidation retards the welding of deformed ball into the bond pad, and stitch into the lead frame or substrate. Today, this has been overcome due to the vast improvement in the wire bonding technology and processing of different materials (e.g. copper wire, lead frame or substrate, device metallization, etc.) to complement the use of copper wire.

- The utilization of forming gas (a mixture of 95% Nitrogen and 5% Hydrogen)- for an oxidation free process during the formation of copper free-air-ball (FAB).

- All automatic wire bonders used for copper wire bonding process are all equipped with copper kit, comprising of EFO (electronic flame-off kit) with provision to ensure optimum flow of forming gas.

- Palladium coated copper (Pd coated Cu) wire is an alternative to choice retard oxidation.

- Software enhancements integrated in the new generation of copper wire bonders to improve ball bondability with minimal aluminum splash-out and programmable segmented stitch features.

- Special type of capillary surface finishing with granulated protrusion for better gripping and to reduce short tail related stitch bondability problems.

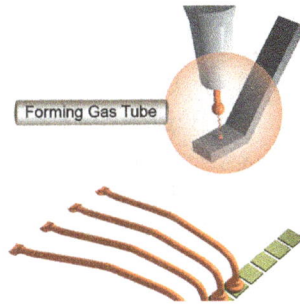

## Wire Bonding Cycle

The wire bonding cycle using copper wire is almost the same as the gold wire as it forms the ball bond, loop, and stitch sequence. The introduction of forming gas during the free-air-ball formation for copper wire is the only difference in the process. Forming gas consist of 95% Nitrogen to prevent copper wire from oxidation and 5% Hydrogen for flammability enhancement to create concentric FAB during EFO (electronic flame off) firing. Highly oxidized copper free-air-balls are basically harder and more difficult to bond on sensitive silicon technology. In addition, the forming gas helps to inhibit the oxidation of copper wire once exposed to ambient temperature of wire bonders' heater block.

## Free Air Ball

To achieve consistency of the free air ball size, it requires consistent tail length after second bond formation, and consistent electronic flame-off (EFO) firing.

## Gold Wire

After the second bond application, the capillary lifts up with its tail protruding outside the capillary tip. This action would then enable the electronic flame off (EFO) to be activated to form the gold FAB and then the bondhead goes to the reset position for the next bonding cycle. Thus, the FAB is already formed before the next bonding cycle begins.

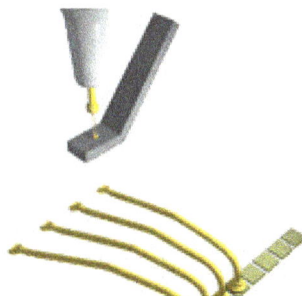

## Copper Wire

To form a concentric copper free-air-ball, a "Copper Kit" is a must have to ensure continuous controlled flow of forming gas towards the tip of the capillary and torch electrode (EFO Wand) to prevent copper from oxidation. After the stitch bond application, the capillary lifts up with its tail protruding outside the capillary tip then the bondhead goes to the reset position with the absence of the FAB. The copper bonding cycle starts with a formation of the FAB.

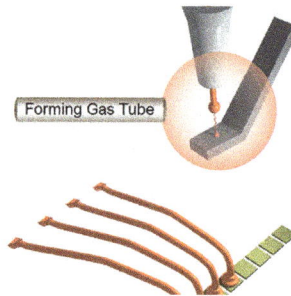

## Free-air Ball is Centered Inside the Capillary Chamfer Area

A wire tensioner is used to ensure that the free air ball is up and at the center of capillary face prior to being lowered onto the die bond area. If this condition is not met, there is a chance of producing an irregular ball bond deformation commonly known as "golf club ball bond".

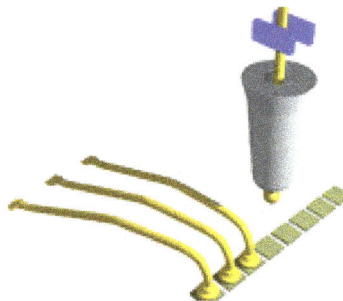

## Ball Bond Formation

The capillary is lowered with the free air ball at its tips' center, and initial ball deformation is made by the application of impact force.

The application of the ultrasonic energy, force, temperature and time enabled the initial ball to be deformed further to the geometrical shape of inside chamfer, chamfer angle and the hole.

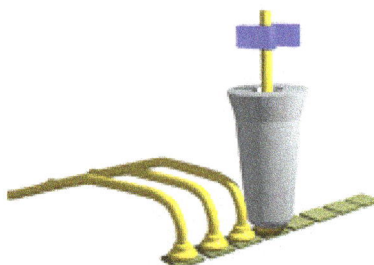

## Capillary Forms the Loop and then Stitch Bond

After the ball bonding, the capillary raises, looping takes place as the capillary travels at the same time from the first position of the ball bond to the direction of the second bond to form the stitch.

The looping can be varied to a different modes depending upon the device/package type. Achieving low-loop, long lead bonding is no more a problem because of the programmable looping algorithm that optimizes its formation for each different lead length.

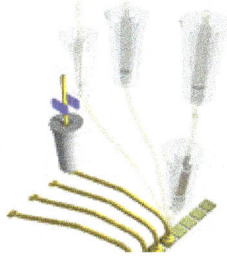

## Tail Formation after the Stitch Bond

Once the capillary reaches the targeted second bond position, the stitch is then formed with similar factors applied during the first bond. The capillary deformed the wire against the lead or substrate producing a wedge-shaped impression.

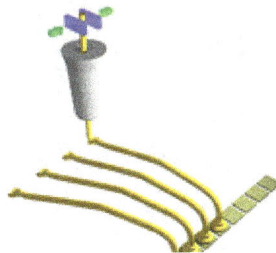

## Capillary Lifts and Forms a Tail

It is important to note that a certain amount of tail bond is left to allow pulling of the wire out of the capillary after the stitch bond application in preparation for the next free-air-ball formation.

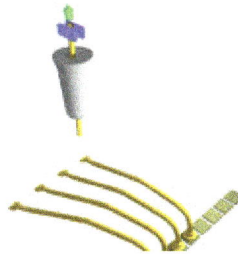

## Wire Bonding Material

Wire bonding materials used in a ball bonding process mainly include the bonding wire and bonding tool. Ball bonding tools are called capillaries, which are axial-symmetric ceramic tools with vertical feed holes. Figure below shows an example of a capillary used in fine-pitch applications. The tool's tip is shaped to give the clearance needed in fine-pitch bonding.

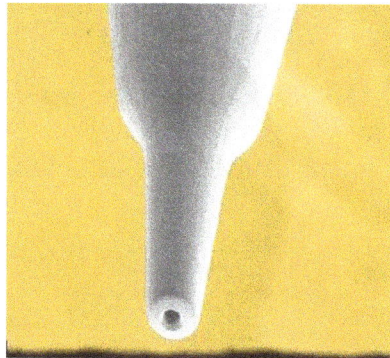

Figure: Wire bonding tools.

Figure below outlines the critical dimensions of a capillary, which include the tip diameter (T), angle of the bottom face (FA), outside radius (OR), hole diameter (H), and chamfer diameter (CD). The tip usually is determined by an application's pitch. FA and OR affect mainly second bond, while the hole and chamfer diameters affect both the first and tail bond formations. These are the most critical dimensions of a capillary.

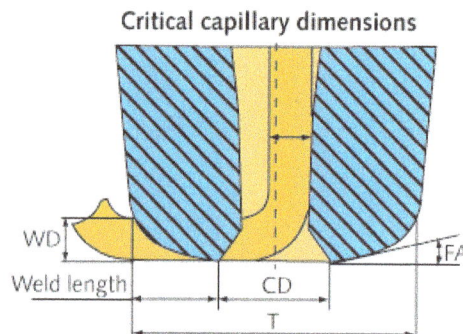

Figure: Critical capillary dimensions.

Most bonding wire used in ball bonding is gold (Au) wire of 99.99% purity, which is often referred to as 4Ns wire. Alloy wires (99.99% or less purity) are sometimes used to meet special application requirements, such as high wire strength. Studies have shown that certain dopant (impurity in the wire) can slow Au-Al intermetallic growth. 3Ns and 2Ns wires are sometimes considered to improve device reliability.

A special consideration for bonding wire is its heat-affected zone length, which is related to the recrystallization process due to the heat from EFO. The heat-affect zone often weakens the wire. A longer heat-affect zone in the wire often results in higher loop height. Some low-loop applications require high strength and a low heat-affected zone, as shown in figure below.

Copper (Cu) wires can be bonded with some modifications to the wire bonder. The modifications mainly consist of using a forming gas environment to prevent Cu oxidation during the free air ball formation. Both Au and Cu bonding are done at an elevated temperature (normally 150° to 240°C, depending on the device). This process is called thermosonic bonding because of the use of heat and ultrasonic energy.

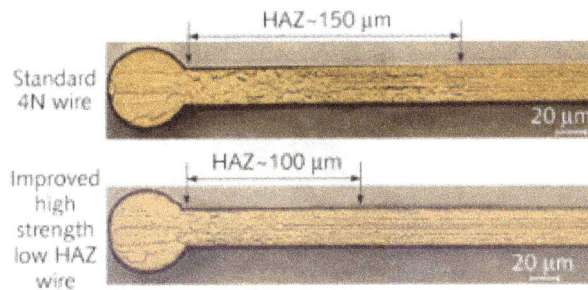

Figure: Certain high-strength wire offers shorter
heat-affected zone (HAZ) and improved looping capability.

## Wedge Bonding

The ultrasonic bonding process typically started by feeding the wire at an angle usually 30-60° from the horizontal bonding surface through a hole in the back of a bonding wedge. Normally, forward bonding is preferred, i.e. the first bond is made to the die and the second is made to the substrate. The reason is that it can be far less susceptible to edge shorts between the wire and die. By descending the wedge onto the IC bond pad, the wire is pinned against the pad surface and an U/S or T/S bond is performed. Next, the wedge rises and executes a motion to create a desired loop shape. At the second bond location, the wedge descends, making a second bond. During the loop formation, the movement of the axis of the bonding wedge feed hole must be aligned with the center line of the first bond, so that the wire can be fed freely through the hole in the wedge. Several methods can be used to end the wire after the second bond. For small wires (<.003"/76μm), clamps can be used to break the wire while machine bonding force is maintained on the second bond (clamp tear), or the clamps remain stationary and the bonding tool raises off the second bond to tear the wire. The clamp tear process offers a slightly higher yield and reliability than the table tear process due to the force maintained on the second bond during the clamp tear motion. The clamp tear process also offers a slight speed advantage over the table tear process due to fewer required table motions. However, the table tear process, has a higher wire feed angle capability and stationary clamp, has the potential to provide slightly more clearance from package obstructions such as a bond shelf or pin grid. For large bonding wires (>.003"/76μm), the most common method is using a cutter blade. Once the wire is terminated, the wedge ascends. The clamped wire is fed under it to begin bonding the next wire. This process will repeat until the wire bond program is complete.

Clamp Tear

Table Tear

Guillotine termination methods

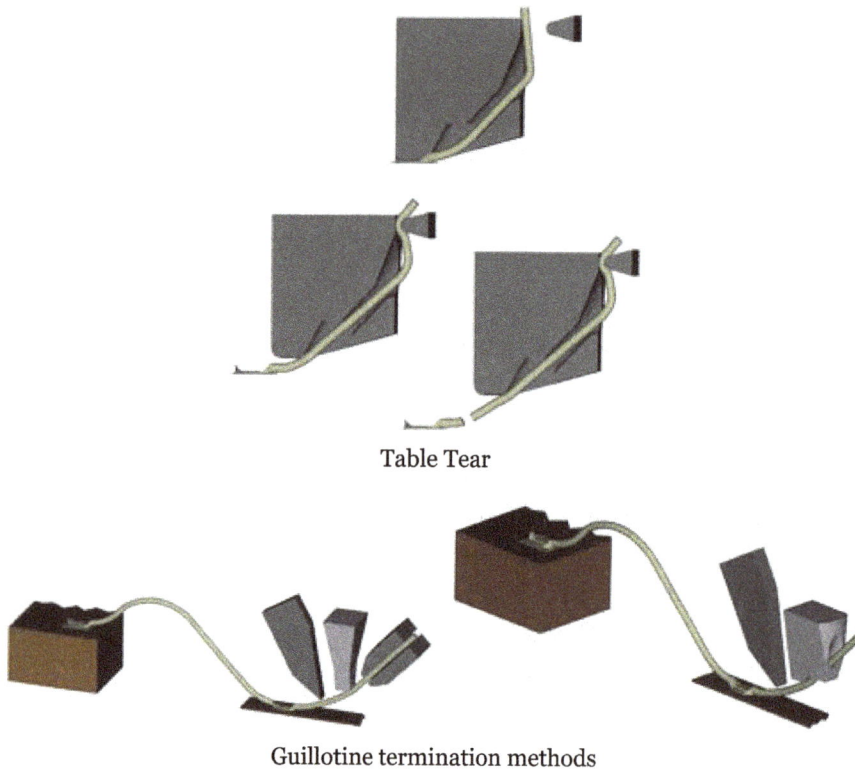

Wedge bonding technique can be used for both aluminum wire and gold wire bonding applications with some slight modification to the back radius to compensate for the lower tensile strength of gold wire. The principle difference between the two processes is that the aluminum wire is bonded in an ultrasonic bonding process at room temperature, whereas gold wire wedge bonding is performed through a thermosonic bonding process with heat up to 175° C. A considerable advantage of the wedge bonding is that it can be designed and manufactured to very small dimensions, down to 50μm pitch. Aluminum ultrasonic bonding is the most common wedge bonding process because of the low cost and can be bonded at room temperature. The main advantage for gold wire wedge bonding is the possibility of avoiding the need for hermetic packaging after bonding due to the inert properties of the gold. In addition, a wedge bond will give a smaller footprint than a ball bond, which specially benefits the microwave devices with small pads that require a gold wire junction down to .0005"/13μm.

## Compliant Bonding

Solid-state joining of metal couples using mechanical and thermal energy has been employed in processes such as forge welding, cladding, and pressure welding. With the emergence of solid-state electronic devices, the need to join metal leads to metallized semiconductor surfaces well below their respective melting points led to the development of thermo compression bonding (figure).1 The bond is formed by inducing a suitable amount of material flow in the lead by a heated bonding ram so that adhesion takes place in the absence of a liquid phase. To extend the material range of solid-state bonding and eliminate the need of applying heat from an external source, a process referred to as ultrasonic bonding was developed. The bond is formed by applying high frequency vibrations from a ram to mating metal surfaces in contact under relatively low pressures.

A significant number of lead failures bonded by thermo compression or ultrasonic techniques are associated with the rigidity of the bonding ram where the cross-sectional area of the lead is excessively reduced. This condition is generally relieved by controlling the bond parameters and contouring the face of the ram (figure). Economic advantages of attaching a multiple number of metal leads simultaneously with a single-energized ram presented additional problems related to the rigidity of the ram. For example, a uniform transmission of energy from the tool to each lead throughout the joining cycle is difficult to attain. This is generally due to the need to critically control the bond parameters, fine misalignments, imperfections (wear or flaws) on the ram surface, variations in lead thickness, and the natural topography of Le substrate. With smaller leads such as those incorporated in beam lead devices, tolerance considerations become more important. And as beam lead spacing's decrease, control of bond energy transmission becomes paramount in order to avoid bridging of adjacent leads.

Compliant bonding substantially reduces problems related to conventional bonding rams, since it incorporates a deformable or compliant medium between an energy source and bond region(s). Due to the inherent flow properties of compliant media, the process markedly reduces tolerance problems, automatically controls the extent of lead deformation, and facilitates transmission of a uniform quantity of bond energy to a multiple number of leads in one cycle.

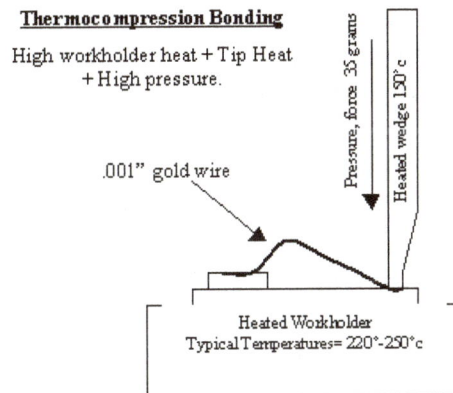

Figure: Illustration showing various means of controlling the bond structure such as contouring the ram (chisel and eyelet2) and "balling" the lead

In contrast to other solid-state bonding processes which are highly equipment dependent, compliant bonding offers a unique concept in joining where the mechanism of energy transmission is controlled by the properties of compliant media. And as a result, the process offers both reliability and a standardization of process variables for both laboratory and manufacturing scale environments.

## Modes of Compliant Bonding

Compliant bonds are formed by compressing deformable or compliant media against the topography of bond regions* with suitable energy sources. The process offers a wide range of flexibility since various energy sources and compliant media may be developed for a ~articular application. For example, bond energy may be supplied to compliant media by thermally (figure below) or ultrasonically activated rams. In this case, compliant media are generally thicker than the lead or chip. Development of thin compliant membranes, less than the thickness of the bond regions and compressed by heated fluids, is expected to produce similar bonds. Combinations of flexible membranes (as energy sources) compressing thick compliant media would facilitate bonding over larger and irregular contoured areas.

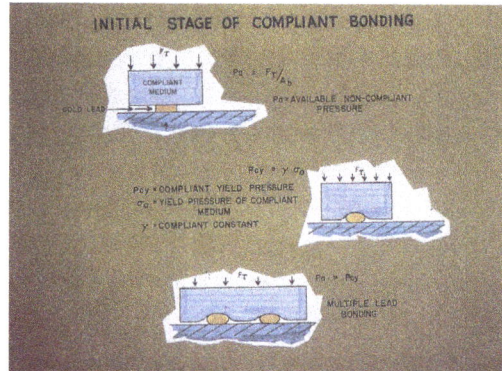

Figure: Illustration showing the stages of compliant bonding gold leads.

Though heated rams (energy source) and relatively thick compliant media have been extensively developed, cursory experiments with ultrasonic activated rams have been performed. For example, metallized silicon devices and gold wires have been compliant-sonic bonded to gold thin films by techniques illustrated in figure. Feasibility in attaching gold wires and conventional silicon chips was demonstrated using compliant media such as five-mil Kapton and one-mil Kapton cladded to five-mil aluminum tape. The gold wires were controlled from excessive deformation by toe flowing media. In attaching conventional silicon devices, the compliant media partially conforms to the device by a slight vertical force of the ram, producing a coupling effect for transmitting the lateral oscillatory motion of the ram to the bond interface. Similar mechanical coupling would require a precision-shaped ram for each device. In some cases, the increased rate of mechanical stress introduced parallel to the medium-coupled chip by high frequency oscillatory motion could result in a stiffer or more efficient transmitting media. This is expected to occur for materials whose flow stress properties increase with strain rates comparable to the high frequency motion. Thus compliant media may be developed which easily conform to the device while transforming to a stiffer material during the transmission of lateral motion. Obviously, the media may also serve as a chip carrier to the bonding station.

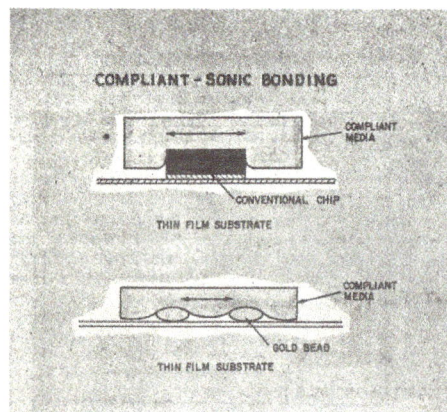

Figure: Illustration showing the transmission of ultrasonic energy through compliant media to a conventional chip and gold leads. The energy is applied by compressing the media with an ultrasonically-activated ram.

## Compliant Bonding Gold Leads

Solid-state bonds between gold leads and gold metallized surfaces may be formed by inducing a suitable amount of material flow at the· interface by the application of heat and pressure so that adhesion takes place in the absence of a liquid phase. Processes utilizing this bonding mechanism are cladding, pressure welding, forge welding, and thermo compression bonding - the latter being widely applied to attaching electrical leads.

## Compliant-bond Structure

Solid-state bonds may be accomplished by compressing a 2024 aluminum alloy (compliant media) against a gold lead with a heated ram (energy source). As the' lead is deformed against the metallized gold surface during the bond cycle, it also plastically penetrates the compliant media. This produces controlled deformation of a compliantly bonded gold wire which is largely attributed to two separate mechanisms associated with the mutual flow of the lead and compliant media. By compressing a 10 mil thick 2024 aluminum compliant tape against a 5 mil gold wire with a flat-faced, heated ram, the gold wire deforms more in the central portion of the bond structure, while the deformation 'gradually decreases near the free-length portion of the wire. The deformation gradient is associated with greater forces transmitted to the wire by the compliant media in the central. Portion as compared to the ends where the media is free to extrude over the wire. Secondly, controlled sidewise flow of the lead is due to the conforming compliant media as it contacts the sub-substrate on either side. Similar to thermo compression bonds formed with contoured rams and critically controlled bond forces, the resultant compliant bond structure effectively eliminates problems associated with excessive lead deformation. The' structure is simply formed with a flat-faced ram and the inherent flow properties of compliant media.

## Compliant Pressure Transmission

To obtain a sufficient amount of pressure transmission with a metal compliant media, their relative flow properties at their respective temperatures must be such that upon compression both .the lead and compliant media deform. Pressures to sufficiently deform the lead against the mating surface and form the indentation of the lead in compliant material are not equal to their respective tensile or compressive values. Stress to deform the bulk lead is generally higher since it is being compressed against a non-lubricated gold surface. Stress to obtain an indentation of the deforming lead in a bulk compliant metal is also higher. And since the compliant media deforms, it becomes the rate controlling step for transmitting the bond pressure to the lead. This results in a self-controlling property unique to compliant bonding. For example, with other solid-state bonding processes, the ram force and bond area define the bond pressure, whereas in compliant bonding with relatively thick compliant media, the applied bond pressure is dependent upon the' flow stress properties of the compliant media.

As a deforming lead penetrates a metal compliant media, stresses are set up in various directions "in the media as opposed to simple tensile or compressive stresses. Thus the mean bond pressure, Pc, developed at the compliant-lead interface is greater than the inherent flow stress properties of the compliant medium.

$$P_c = \gamma\sigma$$

Where

> $P_c$ Compliant bond pressure
>
> $\sigma$ Flow stress of the compliant medium
>
> $\gamma$ Compliant constant $(>1)$

In the case of compliant bonding 'gold leads, the yielding or bond pressure would, of course, be a maximum at the central portion of the lead where the compliant medium is largely constrained. Thus the rate controlling step for compliant flow would be controlled by the flow stress properties of the media in the central portion where the pressure for compliant flow is highest.

Mean pressures developed at the compliant lead interface are analogous to pressures required to plastically penetrate a bulk metal with a hard, spherical indenter, except for the unique condition that the. Penetrator {gold lead) changes shape during the process. According to the Tresca or Huber-Mises criterion, the onset of plastic deformation on a metallic surface by a hard, spherical indenter commences when:

> $P_m = 1.1\sigma_0$

Where

> $p_m$ = mean pressure for the onset of plastic deformation
>
> $\sigma_0$ = yield strength

As the pressure increases, plastic deformation in the vicinity of the indenter increases. until the whole of the metal immediately around the indentation is in a state 6f plasticity while the rest remains largely elastic. This prevents the spread of plastic flow in the bulk metal. The mean pressure increases to a value of about $3\sigma_0$ as the deformation passes from the onset of plastic deformation to the fully plastic state. The compliant constant, $\gamma$ (Equation above), is expected to equal a similar value.

## Line Driver

A device that can use installed twisted-pair phone lines or leased lines to connect terminals to servers in different parts of a building or in different buildings.

A line driver is essentially a combination of a signal converter and an amplifier for digital signals. The signal converter performs line conditioning, and the amplifier increases the signal strength.

Also called a "short-haul" device, a line driver allows a signal produced by a serial transmission device using an interface such as RS-232 to be carried over a longer distance than the interface standard allows, which for RS-232 is only 15 meters.

## How it Works

Line drivers are always used in pairs. One line driver is placed at the local site and is connected to the terminal, while the other is located at the remote site and is connected to the server. Line drivers are typically used to extend the maximum distance of serial communication protocols such as RS-232, V.35, X.21, and G.703 and can provide either synchronous or asynchronous communication in various vendor implementations. Considerations for line driver type include full-duplex or half-duplex communication, 2-wire or 4-wire cabling options, and various kinds of connectors.

The most common type of line driver uses an RS-232 serial interface for synchronous transmission of data over installed 4-wire telephone cabling. These line drivers can extend the maximum distance of RS-232 serial transmission from 15 meters to several kilometers.

For intrabuilding connections using line drivers, copper unshielded twisted-pair cabling or the installed telephone lines are typically used. For interbuilding connections, fiber-optic cabling is preferred.

Line drivers are available for almost every kind of communication mode, from 19.2-Kbps RS-232 serial line drivers over 6 kilometers to 2-Mbps single-mode fiber-optic line drivers over 18 kilometers. Line drivers for parallel connections can extend parallel transmission of data from about 6 meters to several kilometers. Line drivers are also used in implementation of $T_1$ lines.

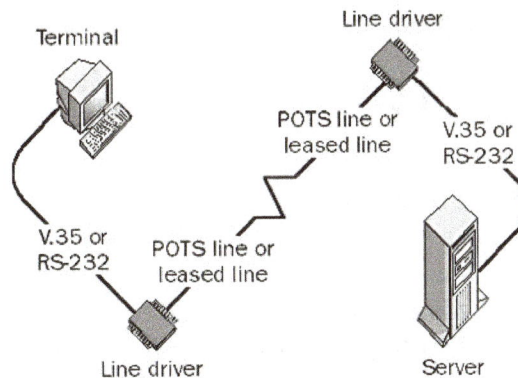

When you use line drivers, your maximum bandwidth and transmission distance are inversely related - that is, the longer the line, the less bandwidth you have.

## References

- Floorplanning: vlsibyjim.blogspot.com, Retrieved 31 March 2018

- The-basics-of-wire-bonding: neotech.com, Retrieved 16 June 2018

- Wire-bonding-tutorial: electroiq.com, Retrieved 19 May 2018

- Basic-ultrasonic-wedge-bonding-process, technical-overview, wedge-bonding-tools, chip-bonding-tools, products-and-solutions: smallprecisiontools.com, Retrieved 19 July 2018

- Line-driver: thenetworkencyclopedia.com, Retrieved 29 June 2018

# 5

# VLSI Architecture

An understanding of VLSI architecture requires comprehension of the fundamentals of hardware description language, register-transfer level, pipeline in computing and globally asynchronous locally synchronous architecture. This chapter covers all these essential aspects for a thorough understanding.

Digital systems have traditionally used a variety of technologies and manufacturing processes. The engineering of systems composed of many parts allows for a degree of specialization of the technology to the function to be performed, be it logic, storage, or communication. The progress in microelectronics, particularly in the last decade, has changed this engineering situation dramatically. As the level of integration has increased, and the number of manufactured parts required to accomplish a given function has decreased, the engineering situation has become more uniform. Most of the design and engineering effort for a digital system today occurs in the design of the chips. Once the attention is inside a chip, the technology with which one creates a system is very tidy and consistent, governed by a much smaller set of rules and design paradigms than when many technologies have to be considered, and with no practical "escape" into other technologies to bail the designer out of a tough problem.

## VLSI Models of Computation

A consequence of this uniformity is that VLSI models of computation are quite realistic as a means of quantifying the consequences in silicon area, a measure of-cost and computing time, of architectural choices within a chip. When a variety of digital technologies had to be considered, each with its- own cost, performance, and functional specialization, such modeling was much less tractable, or could be carried out only at a coarse level. As the scale of a system or subsystem encompassed in a single chip increases, one might hope and expect that architectural tradeoffs will be accomplished in a less ad hoc fashion. VLSI is indeed a beautiful medium for studying the structure and design of digital systems, and this fact, as much as its economic importance, explains its appeal in the research community.

VLSI models of computation have been used extensively for the complexity analysis of concurrent algorithms. Although they are abstractions from the actual complications in chip design, they provide a way to describe concisely several of the essential features and problems of VLSI design that lead one toward concurrent architectures. We require, and hence shall use, only rather crude models here. For area A, it will suffice to use the actual or approximate area on silicon, an existent upper bound. The scalable parameter A is used as the linear unit, where $2\lambda$ is the feature size of the process, so that the area in $\lambda^2$ units is actually a measure of complexity. The computing time $T$ for an element or system is taken as the latency, the delay from input to output for a "problem instance." Thompson suggests that a system working on p problem instances concurrently should be taken to exhibit an area A that is its total area divided by $p$.

For our simpler purposes we shall instead take A simply as area, and $T/p$ as the average interval between problem instances at the input and output. Average throughput is then $p/T$, and cost/performance corresponds to $AT/p$.

Concurrency may be exploited at any level of a VLSI system design. One of the most common strategies in the logic, organization, and architecture of VLSI systems is to use pipelines - intermediate storage in computation and communication paths, in order to increase throughput even if it is at the expense of increased latency. Figure is a simple illustration of this approach, the evaluation of an expression (in infix notation) $\big(\big(\big(A\,op_1 B\big)op_2 C\big)op_3 D\big)$, where the $op_i$ are binary operators. Although the operations must be performed sequentially, this evaluation is assumed to be required repetitively. The boxes indicate the temporary storage for the input and output. If this picture were a concurrent program schema, the circles would represent processes, the boxes their input and output queues and the arcs the message paths. If it were a block diagram, the circles would represent combinational functions, the boxes registers, and the arcs bundles of wires. If it were a logic gate diagram, the circles would be gates, the boxes clocked storage elements, and the arcs wires. If one denotes the respective areas and times of the operators as $a_i$ and $t_i$, neglecting the boxes, the areas and times simply accumulate, so that the cost/performance is just $(a_1 + a_2 + a_3)(t_1 + t_2 + t_3)$.

A pipelined version of this system, shown in figure below, is dealing with three problem instances concurrently. Here one sees in miniature some of the opportunities and problems one faces with concurrent computations. If the operation times were equal, one has a system with only slightly increased area (the boxes), the same communication plan, the same latency time, and with $p = 3$, three times the throughput. What if the $t_i$ were not equal? Clearly the throughput would become $1/t_{max}$. When such a system is designed in silicon, one can generally trade off area and time within the operators, or locate the pipeline synchronization, to make the times approximately equal. When this situation appears in programming concurrent computers, it is referred to as load balancing, and is dealt with in ways that depend on both the application and on the architecture.

Although there is apparent localization of communication within and between the operators, this example is otherwise independent of technology considerations. The case in favor of concurrency in VLSI systems becomes more compelling when one examines the area-time performance of the communication.

There are currently in vogue several different time models for the delay of a wire and its driver. Each is physically valid under a range of conditions in a particular technology. In MOS technologies a simple amplifier with input capacitance $C_{in}$, will drive a capacitive load $C_{out}$ in time $T = \tau_{inv}\big(C_{out}/C_{in}\big)$, where $\tau_{inv}$ is a characteristic of the process, essentially the delay of an inverter with unit fanout and no parasitic load., The area of the amplifier is proportional to $C_{in}$. By cascading amplifiers in an optimal size ratio $C_{out}/C_{in} = e$, one can boost a signal from an energy corresponding to capacitance $C_x$, to the parasitic capacitance of a wire, $C_L$, in a minimum time $T = \tau_{inv}\log_e\big(C_L/C_x\big)$.

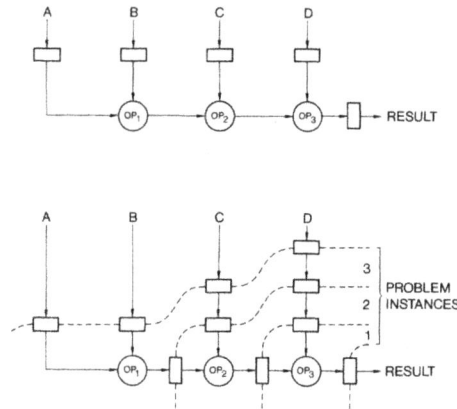

Figure: (a) Cascade evaluation. (b) Pipeline evaluation.

The parasitic capacitance of a wire varies with wire length $l$ according to a proportionality constant of the layer, so one can assert that communication time is $T = O(\log l)$ in the worst case in which one starts from minimum signal energy. The numbers for a typical CMOS process of today are ' $\tau_{inv} = 0.6\,ns$, , and the ratio between the capacitance of largest wiring structures and a minimum size transistor gate is about $e^8$, and scales to about $e^{15}$ in ultimate MOS technologies. Although in typical practice the compromise between driver area and delay dictates a suboptimal driver for long wires or interchip wires, one can, if necessary, achieve $O(\log l)$ dependence of the driver delay with wire length.

The resistance of a wire together with the parasitic capacitance adds a diffusive propagation term for the wire itself that is $O(l^2)$ . The coefficient of this term makes this phenomenon significant in the fairly resistive silicon and polycrystalline silicon wires today, and is an important performance consideration in memory and programmed logic array (PLA) structures in single-level metal processes, since silicon wires must be used in one of the two directions in the matrix layout. In scaled technologies this problem appears also in metal wires, due to their scaled thickness. For reasons other than performance, such as noise immunity, one would never allow this diffusive term to become seriously dominant. Instead, one can include active repeater amplifiers periodically along a long communication path, thus making the delay $O(l)$ , or one can convey signals long distances on additional metal layers that are thicker.

While the dependence of delay time on wire length is quite benign if one uses optimal driving structures, the combination of area cost and this slowly growing delay is a substantial incentive to localization of communication. Communication throughput, or bandwidth, in the VLSI medium is a very meaningful measure that may be taken as a product of the number of parallel wires (the width of the wiring track) between two parts, and the "bit" rate. In signaling schemes in which there is no pipelining along the communication path, there can be only one transition propagating along the path, and the bit rate is bound by the reciprocal of the delay. If one takes the area-time product as the objective function to be minimized in a particular design, one sees that VLSI imposes serious penalties for separating two parts that communicate in this naive way.

The wire area, the signal energy, and the area of the optimal driver (about 10 percent of the wire area) is each proportional to the distance, that is, $A = O(l)$, while $T = O(\log l)$. Accordingly, the aggregate penalty in cost/performance for violating the principle of locality is $O(\log l)$. The penalty may be still worse for wires that are so long that they must be forced onto the upper, thicker metal layers, since these layers will have coarser design rules, exclusively long wires, and hence more area per wire, and are expected to be a very limited resource.

What one sees in the example of pipelining, and from the area-time performance of communication, is that the throughput can be increased if one can devise a way to confine the (physical) diameter of tightly coupled parts of a system. The expression "tightly coupled" can be taken in the synchronous design style to describe a system in which a large proportion of the communication must traverse the entire system on each clock period, such as from the input to the output registers.

On the other hand, when large systems are composed in a loosely coupled fashion, by which we mean that the parts can operate relatively independently that is, concurrently - and can tolerate latency in their communication with other parts, the raw performance and excellent cost/performance that can be achieved in small diameter systems is reflected in large systems. This principle has been applied so thoroughly in the examples discussed later that these systems are arbitrarily extensible in the number of concurrent computing elements, and open-ended in performance. They can be expanded to be as large as desired with each part still operating at the same rate as when it is incorporated into a smaller system. This property is closely associated with the ability of VLSI architecture to be scaled, that is, to exploit advances in the circuit technology.

Another conclusion of this informal analysis is that communication between concurrent computing elements may take only certain forms. Any wide path -many parallel wires -must be strictly localized, such as can be achieved in a mesh-connected system of concurrent computing elements. Wiring economics dictate that any long path be narrow, and will necessarily exhibit a significant latency. This latency will be due in part to pipelining in communication. Also, messages of many bits sent between concurrent computing elements must be serialized according to the width of the communication path, and would accordingly exhibit a latency that is dependent on the size of the communication "packet". These two communication paradigms -local and wide versus distant and latent- are indeed recurrent and competing themes in the later examples; the former representing the wave front or systolic type of computation, in which all communication can be made local, and the latter the queued message routing approach to less regular computations.

## Architectures that Scale

Another aspect of this increasing uniformity of digital technology is that the future of at least the silicon technology is believed to be well understood and, but for a few considerations, is not radically different from today's high complexity MOS technologies. Thus, one has a reasonable hope of devising architectures with a longevity that parallels the continued evolution of the technology, or in VLSI jargon, "architectures that scale."

The physical consequences of feature size scaling of MOS technologies are generally advantageous. If all physical dimensions and voltages are reduced together , so that electric field strength remains constant (and no new materials need be postulated), the electron "velocity" , $\mu E$ , where , $\mu$ is the mobility and $E$ the electric field, remains constant, and the transit time across the smaller dimension channel is reduced in direct proportion. Capacitances per unit area increase linearly in the scaling, but the areas decrease quadratically, so that the capacitances in a scaled circuit are linearly smaller. One can see that at each circuit node the relation $i\Delta t = \Delta q = C\Delta v$ is satisfied in such a way that the scaled circuit is an exact current-, voltage-, and time-scaled replica of the original.

The energy associated with each switching event, $(1/2)C\Delta v^2$, is scaling as the third power of the feature size. If $C$ is taken as the capacitance of the gate of a minimum geometry transistor, this switching energy $E_{sw}$, is equivalent to the product of the power per device and device delay, with the device switching at maximum rate. The power-delay product is the fundamental figure of merit for switching devices, and has a direct relation to the cost/performance of systems implemented with those devices. One notes also that the power per device scales down quadratically while the density scales up quadratically, so that the power per unit area need not be increased over today's levels.

This cube law scaling of $E_{sw}$ with feature size is a remarkable incentive for continued advances in MOS technologies. If one were to take an existing single chip in a 2 $\mu m$ CMOS technology, say, an instruction processor about 5 mm square running at a 20 MHz clock rate, and fabricate it at one-tenth of its present feature size, it would take up only $\frac{1}{100}$ the area, and in principle would operate from an 0.5 V source at $\frac{1}{100}$ the power, and at a 200 MHz clock rate. This is certainly an attractive scaling. In practice, there are a few things that go wrong in this scaling that would have some impact on the scaled performance, but they are only difficulties, not disasters, for computing elements that are the complexity of a single chip today.

Operation at such small voltages is problematic, and due to short channel effects, the threshold voltages and performance of the transistors would not be quite as good as simple scaling rules imply. Another effect that is significant is that while the capacitance distributed along the wires and transistor gates of a circuit node would be scaled to $\frac{1}{10}$ the previous value, the reduced cross section of the wires would cause the resistance of the wire segments, even though they are shorter, to increase by a factor of 10. The RC product, which has dimensions of time, and is the coefficient in the delay in diffusive propagation of signals on wires, unfortunately fails to scale with the other circuit delays. Also, unless temperature is also scaled, the sub threshold conductance of the MOS transistor scales up rather dramatically with reduced threshold voltages, so that the dynamic storage structures that are so widely employed in MOS technologies, such as for dynamic RAM's, cannot be depended on to retain charge for more than a few microseconds. One can get around each of these scaling difficulties by variations in design style. For example, the use of additional metal layers in layout would be a compensation for the problem of wire resistance, and one would expect increased use of static storage structures in place of dynamic storage.

So, it appears that variations of the designs of today, and of the design techniques, are feasible at least in small areas of the chips in this futuristic technology. The design of a chip with an area similar to today's chips, but with 100 times higher circuit density, will depend strongly on the additional layers of metal interconnection mentioned above. This situation is not at all unlike the wiring hierarchy employed today with packaged chips on printed circuit boards or chips mounted directly in ceramic carrier modules, but with the next level of wiring absorbed into the chip. Even under the optimistic and uniformitarian view that this future technology inherits our present design practices nearly intact, there is no fundamental help in sight for relieving the communication limitations. Indeed, by confining the wiring to two dimensions, we give up a physical dimension of packaging and interconnection. Driving a signal that is equivalent to a cross-chip wire of today has become no easier in scaling, and because of the diffusive delays in the lowest levels of interconnect, will force some connections to higher levels. Driving the long distance interconnect on the upper metal layers is essentially similar to driving bonding pads, package pins, and interchip wiring today.

The concurrent VLSI architectures were selected as examples based on being 1) systems for which at least prototypes exist, and 2) clearly inspired by the opportunities and consistent with the design principles of VLSI. This selection is a family of concurrent systems that have been referred to as ensembles, and which can be discussed in terms of a simple process model of computation. There are other computational models and concurrent architectures whose VLSI implementations are interesting, such as data-flow and reduction machines, but these subjects would entail a survey in themselves.

There is a commonality in the physical structure of this family of systems that is in part a necessity of the VLSI medium, and in part an artifact that we shall try to identify, these examples are all regularly connected direct networks of nominally identical concurrent computing elements. For the following discussion and examples, we shall refer to the computing elements as nodes, as in a computer network, and the, communication paths between them as channels.

## Communication Plans

The communication plans of these systems are direct networks, such as the diverse selection. In some systems the communication plan is a direct mapping of the communication requirements of an algorithm. In other cases messages are routed to a destination node through intermediate nodes, and the choice of communication plan is a compromise between wirability and performance. For example, a family of hypertorus networks can be represented as a $k$-coordinate periodic $n$-dimensional cube ($k$-ary $n$-cube) that connects $k^n = N$ nodes together such that the maximal shortest path between nodes is $kn/2$. In order to connect $N = 2^{12}$ nodes, one might choose $k = 2^6$ and $n = 2$, an easily wirable two-dimensional mesh with $2 \times 2^{12}$ short channels, for which $kn/2 = 64$; or $k = 2$ and $n = 12$, for a binary (or Boolean) 12-cube with $6 \times 2^{12}$ channels, 1/6 of which are as long as a radius of the system, for which $kn/2 = 12$; or some intermediate compromise. Similarly, there are many parametric variations of the binary $n$-cube connected $m$-cycle for connecting $m2^n$ nodes. The width of the communication path is still another engineering variable.

In the VLSI engines discussed here, the nodes themselves produce, consume, and in some cases route messages, so that these are what are called direct networks. In systems in which messages are routed, the node could be partitioned into two concurrently operating sections, as illustrated in figure below (a), one section ($C$) to compute, and the other section ($R$) to route messages. The network illustrated is a binary 3-cube, and the channels are labeled according to the dimensions 0, 1, 2. The message section can be reorganized by the transformation shown into the multistage routing network of interchange boxes. This transformation of the direct Boolean n-cube illustrates its essential similarity in structure and message flow performance to the corresponding indirect network, which is the same as the "flip" network used in STARAN, and under a rearrangement the same as the Omega network or the banyan. The absence of indirect networks in current experiments with concurrent VLSI systems is probably partly an artifact of the VLSI "lore" that switching networks do not scale well, which is certainly true of some of them, such as crossbar switches. Message switches are not eliminated by direct networks, but rather are partly concealed in this fully distributed form.

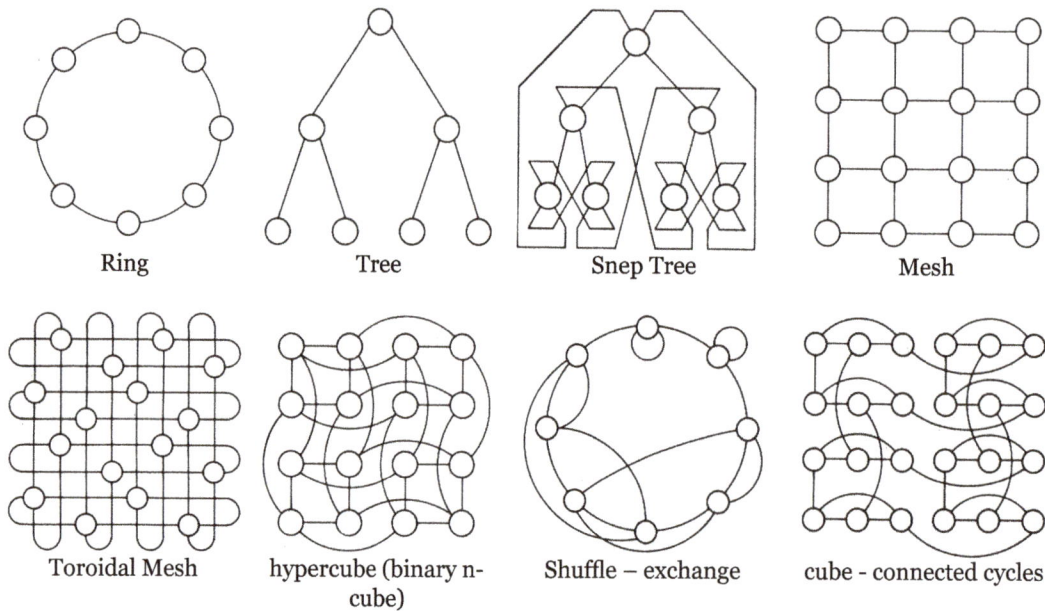

Figure: Typical direct networks

Ring    Tree    Snep Tree    Mesh

Toroidal Mesh    hypercube (binary n-cube)    Shuffle – exchange    cube - connected cycles

## Homogeneity

Another common characteristic of these experiments with concurrent VLSI architectures is that the nodes are nominally identical. We accordingly refer to these machines as homogeneous, meaning that they are of uniform structure. Heterogeneous systems would allow nodes to be specialized for different functions, much as are different computers on a network, or the functional elements in high performance computers. For these early experiments, however, homogeneous machines are certainly logistically simpler to design, test, assemble, and maintain. Homogeneity in programmable machines simplifies the software by giving all parts the same capabilities. Homogeneous machines also conform very well to the design flow of VLSI chips, in which repetition and regularity simplify the layout process, and to the economies of relatively larger fabrication runs of a smaller set of chip types.

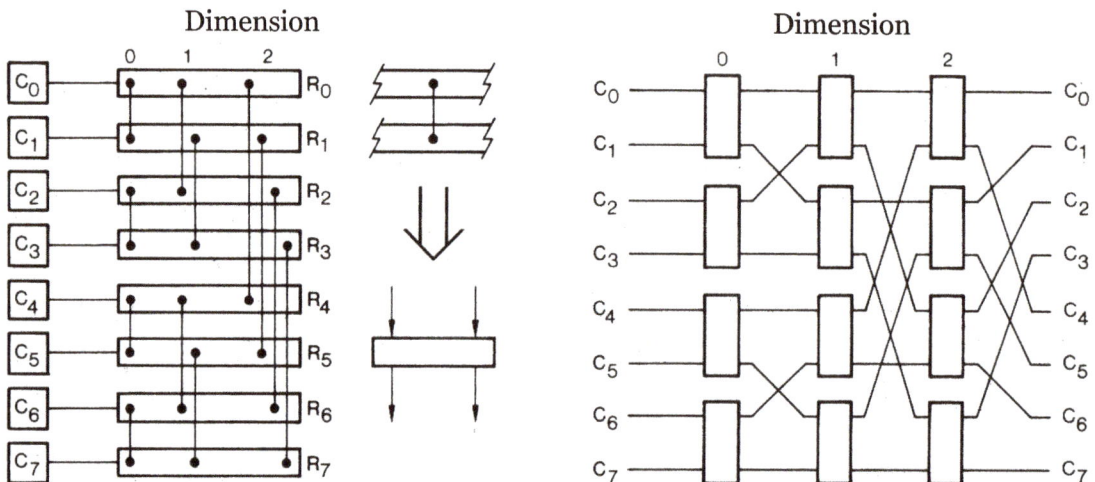

Figure: (a) Direct binary 3-cube, and a transformation from the direct connection to interchange boxes. (b) Indirect binary 3-cube of interchange boxes.

## Node Complexity

The choice of connection network establishes one dimension of variation in taxonomy of this family of concurrent machines. Two other interesting and discriminative dimensions are the complexity of the nodes and the number of nodes reasonably contemplated for a given machine.

The node complexity, also referred to as the "grain size" of the system, appears on the horizontal axis in $\lambda$ area units. The complexity of today's single chip, marked by a "∗" on this axis, is an interesting point that separates systems of many nodes per chip from those of many chips per node. Today's "commodity" chips routinely reach 5 mm on a side at 2.5 $\mu m$ feature size, or 4000 on a side, which translates to 16 $M\lambda$, while advanced commercial chips are in the 50-100 $M\lambda$ range. Of course, these measures have been doubling approximately every two years over the recent past.

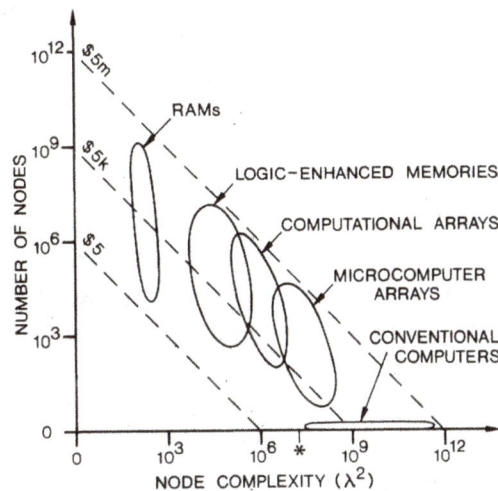

Figure: Taxonomy of concurrent VLSI systems

The two extreme zones indicated in figure above, storage systems composed of random-access memory (RAM) chips, and conventional single processor computers, are included for comparison with the three middle zones. For this comparison, we take the RAM cell and whole computer as the "node." The three middle zones represent concurrent VLSI engines whose design, engineering, and applications have been studied in some depth, and which appear to be reasonably distinct classes. The examples in the three following sections are respectively what are labeled as logic-enhanced memories, computational arrays, and microcomputer arrays. Let us here briefly traverse all five zones, left to right, to describe some of the characteristics of each of these classes.

The RAM systems are composed almost exclusively of high complexity chips, and provide a way to gauge the cost of a system if it were so highly integrated and produced in such large quantities. The basic repeated cell for storing one bit in a RAM varies from about 100 $\lambda^2$ for the densest one transistor dynamic RAM chips to about 400 $\lambda^2$ for high performance static RAM's. Multichip assemblies of many identical RAM chips, the larger capacity systems being typically denser and slower, are accordingly shown as an elongated and slightly slanted zone of expected variation. The selling price of mainframe add-in storage based on *64K* dynamic RAM chips is currently somewhat less than $20 per RAM chip, packaged and powered. These *64K* RAM chips are about 10 $M\lambda^2$. The resulting estimate of $2/ $M\lambda^2$ is too low for chips produced in small quantities, so we will use a more conservative estimate of $\$5/M\lambda^2$ for today's technology, with the understanding that this measure scales down with improvements in the circuit and packaging technologies. Hyperbolas of constant cost, the product of the cost per node and number of nodes, appear in the log-log plot of figure above as straight lines labeled with a cost that applies to the highly integrated implementations that one hopes to achieve with VLSI.

Logic-enhanced memories, also called "smart memories," are very fine grain systems in which each node contains a few to a few hundred bits of storage associated with logic that can operate on the storage contents and communicate with other nodes. Useful systems would include thousands, even millions, of these nodes. The ability to mix logic and storage economically at a fine grain in a single technology, which was not so attractive in earlier digital technologies, is part of what makes these architectures unique to VLSI. Typical applications of these systems are computations with images, such as scan conversion, correlation, and path finding; or database operations such as sorting, association, and property inheritance. There is no real theory or computational model behind the design of these systems. They tend to be in the nature of specialized individual inventions.

The next zone represents systems for highly concurrent numerical computations, in which the nodes are capable of operations such as multiplication and addition, and are connected in regular patterns that match the flow of data in the computation. These computational arrays, also called systolic arrays because of the rhythmic pumping of data in pipelines, can be implemented in a variety of forms. The range of node complexities assumes that the sequencing of operations is either built into the nodes, or that the node responds to control signals that are broadcast into the array in the style of single instruction multiple data (SIMD) machines. However, the systolic algorithms designed for such machines are also highly efficient concurrent formulations for microcomputer arrays. Thus, the computational or systolic array is both an architecture and a computational model, and has stimulated a broad research effort in the design of concurrent algorithms for applications such as signal processing, matrix and graph computations, and sorting.

It requires only several $M\lambda^2$, not even a full chip, to implement a minimal instruction processor and a small amount of storage for program and data. A single chip today is sufficient for a processor with a rich instruction set and several thousand bytes of storage. These highly integrated computers exhibit excellent cost/performance, but the performance and storage comes in fairly small units. Thus, it seems inevitable that people are learning to team up myriads of these computers that fit in units approximating one per chip, or many per wafer, to attack demanding computational problems. Each microcomputer is fitted with a number of communication ports, and the array of nodes is connected in a direct network that is, as usual, dictated either by the application or by message routing and flow performance considerations. Whether one can multiply the computing performance in concurrent execution by the number of nodes, or nearly so, is very much dependent on the problem. All of the concurrent formulations for finer grain machines can be mapped very efficiently onto microcomputer arrays. In addition, machines of this class appear to be capable of performing many of the same scientific and engineering computations for which people today use high-performance vector computers.

The last zone, consisting of conventional single- or several-processor computers, exists in a broad range from the single-chip computer to high-speed supercomputers.

# Hardware Description Language

Throughout the decades of digital computer history, various notations have been developed for capturing the logical behavior of digital circuits at different levels of abstraction. The primary purpose of these notations has been to deemphasize the electronic and fabrication aspects of the design and therefore facilitate documentation, analysis and synthesis of large complex systems. Schematics, Boolean equations, timing charts, state transition tables, block diagrams, and hardware description languages are examples of such notations. Although most of these approaches are being used in practice today, hardware description languages have gained significant acceptance.

A Hardware Description Language (HDL henceforth) is a set of notations, similar to software programming languages, used for modeling the logical function of digital circuits and systems. Compared to alternate forms of design capture, it has been shown, in practice that the use of HDLs shortens the design cycle and yields more robust realizations. Many concede that without HDLs, the design of today's complex circuits would not be possible in a reasonable amount of time. A hardware description can serve as a principal means of communication between members of a design team. The conciseness and readability of HDLs minimize the need for any natural language, and more error prone, discussion of the design. Furthermore, a hardware description can be used as an input to a variety of analysis and synthesis tools. These tools greatly facilitate the verification and realization of the described circuits. Similar to a software programming language where the target machine code is hidden from the programmer, HDLs are independent of any particular target circuit technology. This feature contributes to improved readability and design management. It should be noted that an HDL is a language and only a language and has no apparent algebraic structure in terms of guiding the user to a minimal implementation. However, with practice, the designer will readily arrive at modeling techniques that yield more efficient realizations.

## Levels of Abstraction

Levels of abstraction in a hardware description refers to the amount of design decisions that have been specified in the description of the circuit; The fewer the design decisions, the higher the abstraction level. Three distinct levels of abstraction have been recognized in hardware descriptions: Gate Level (or structural), Register Transfer Level (also known as, RTL or dataflow), and Behavioral Level.

The lowest and most detailed level of abstraction is the gate level where specific gates and components and their interconnections are specified in the description. A gate level model is very specific in terms of the design architecture and what gates are to be used for the final implementation. The corresponding function of the hardware is not evident from such descriptions unless the function of each constituent component is known. One can think of a gate level description as a textual representation of the circuit schematic.

Register transfer, or dataflow, models are more abstract and describe a controlled flow of data between buses and registers in the design. Unlike the gate level models, no specific components are specified in dataflow models since that decision is abstracted out in the interest of efficiency, simplicity, and better design management. Even though the mapping between a dataflow description and a hardware architecture is one to many, the hardware correspondence of each statement is well defined. Furthermore, certain architectural decisions, such as the scheduling of data transfers in relation to the system clocks, are specified in the dataflow models.

The behavioral level is the most abstract level and is used to describe the function of the design without providing any implementation details. Furthermore, hardware correspondence is not well defined and in most cases is not evident from the model description. Behavioral models present an input to output mapping according to the data sheet specification and can serve as a concise documentation medium. Compared to gate level and dataflow counterparts, behavioral models often require less effort to develop and are useful for simulation and functional analysis during early stages of the design.

The motivation for recognizing various levels of design abstraction is to provide ways to manage and analyze any given design at all stages of the design cycle. Top-down design methodologies often start by first contriving more abstract models and then proceed to more detailed descriptions. Bottom-up design approaches often start by evaluating existing models and composing them hierarchically with new models in order to achieve the desired functionality.

## Fundamental Characteristics of a Description Language

A general hardware description language should support all levels of abstraction and a number of other characteristics listed below:

- Computations in hardware occur in parallel and this implies that an HDL should support concurrency or description of parallel actions. Additionally, procedural constructs should be supported in order to facilitate description of hardware algorithms that occur within a specific time unit. These procedural constructs are similar to software languages and improve readability.

- Carriers and operators are an integral part of any HDL. A carrier is either a wire or memory, depending on the latency of information stored in them. Operators take their operands from carriers, perform their function, and return the computed outputs onto carriers.

- HDLs should support a varsity of data types that a carrier can assume. However, the most elementary type is the binary digit or bit. Aggregate types are formed by hierarchically grouping bits into vectors and arrays. Often times operators are applied to arrays or vectors of bits and therefore any viable HDL should provide support for such types and operations.

- Partitioning a large design based on functional coherency, and connecting these partitions hierarchically is common practice among designers. Such approach improves design management and allows design reuse. Therefore, HDLs should support component instantiations and hierarchical configuration of designs. Moreover, parameterization of modules in the context of hierarchical design allows description of generic modules which can be customized when instantiated.

- Another commonly desired feature is the ability to express the dichotomy between data and control. Since different optimization techniques are applied to data and control sections, such separation provides the needed design control.

- Hardware independence is a notable characteristic of any HDL. The language notations should be independent of any specific circuit technology.

## VHSIC Hardware Description Language (VHDL)

VHDL is one of the commonly used Hardware Description Languages (HDL) in digital circuit design. VHDL stands for VHSIC Hardware Description Language. In turn, VHSIC stands for Very-High-Speed Integrated Circuit.

VHDL was initiated by the US Department of Defense around 1981. The cooperation of companies such as IBM and Texas Instruments led to the release of VHDL's first version in 1985. Xilinx, which invented the first FPGA in 1984, soon supported VHDL in its products. Since then, VHDL has evolved into a mature language in digital circuit design, simulation, and synthesis.

## Structure of VHDL

Let's consider a simple digital circuit as shown in figure below.

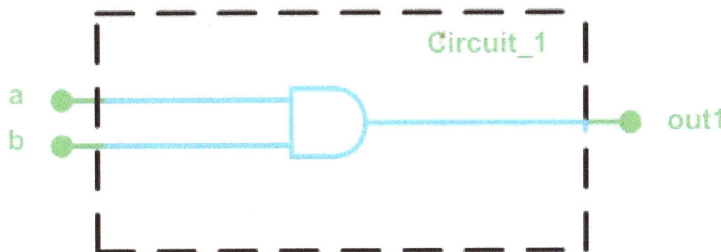

Figure: A simple digital circuit.

This figure shows that there are two input ports, a and b, and one output port, out. The figure suggests that the input and output ports are one bit wide. The functionality of the circuit is to AND the two inputs and put the result on the output port.

VHDL uses a similar description; however, it has its own syntax. For example, it uses the following lines of code to describe the input and output ports of this circuit:

```
1    entity circuit_1 is
2       Port ( a : in  STD_LOGIC;
3              b : in  STD_LOGIC;
4              out1 : out  STD_LOGIC);
5    end circuit_1;
```

Line by line explanation of the above code.

Line 1: The first line of the code specifies an arbitrary name for the circuit to be described. The word "circuit_1", which comes between the keywords "entity" and "is".

Lines 2 to 4: These lines specify the input and output ports of the circuit. Comparing these lines to the circuit of figure, we see that the ports of the circuit along with their features are listed after the keyword "port". For example, line 3 says that we have a port called "b". This port is an input, as indicated by the keyword "in" after the colon.

What does the keyword "std_logic" specify? std_logic is a commonly used data type in VHDL. It can be used to describe a one-bit digital signal. Since all of the input/output ports in figure will transfer a one or a zero, we can use the std_logic data type for these ports.

Line 5: This line determines the end of the "entity" statement.

Hence, the entity part of the code specifies 1) the name of the circuit to be described and 2) the ports of the circuit along with their characteristics, namely, input/output and the data type to be transferred by these ports. The entity part of the code actually describes the interface of a module with its surrounding environment. The features of the above circuit which are specified by the discussed "entity" statement are shown in green in figure.

In addition to the interface of a circuit with its environment, we need to describe the functionality of the circuit. In figure, the functionality of the circuit is to AND the two inputs and put the result on the output port. To describe the operation of the circuit, VHDL adds an "architecture" section and relates it to circuit_1 defined by the entity statement. The VHDL code describing the architecture of this circuit will be

```
6     architecture Behavioral of circuit_1 is

8     begin

9          out1 <= (a and b);

10    end Behavioral;
```

Line 6: This line gives a name, "Behavioral", for the architecture that will be described in the next lines. This name comes between the keywords "architecture" and "of". It also relates this architecture to "circuit_1". In other words, this architecture will describe the operation of "circuit_1".

Line 8: This specifies the beginning of the architecture description.

Line 9 Line 9 uses the syntax of VHDL to describe the circuit's operation. The AND of the two inputs a and b is found within the parentheses, and the result is assigned to the output port using the assignment operator "<=".

Line 10 This specifies the end of the architecture description. As mentioned above, these lines of code describe the circuit's internal operation which, here, is a simple AND gate.

Putting together what we have discussed so far, we are almost done with describing "Circuit_1" in VHDL. We obtain the following code:

```
1     entity circuit_1 is

2          Port (a : in STD_LOGIC;

3               b : in  STD_LOGIC;
```

```
4                    out1 : out  STD_LOGIC);

5     end circuit_1;

----------------------------------------------------

6     architecture Behavioral of circuit_1 is

8     begin

9           out1 <= ( a and b );

10    end Behavioral;
```

However, we still need to add a few more lines of code. These lines will add a library that contains some important definitions, including the definition of data types and operators. A library may consist of several packages. We will have to make the required package(s) of a given library visible to the design.

Since the above example uses the data type "std_logic", we need to add the package "std_log-ic_1164" from "ieee" library to the code. Note that the logical operators for the std_logic data type are also defined in the "std_logic_1164" package—otherwise we would have to make the corresponding package visible to the code. The final code will be

```
1     library ieee;

2     use ieee.std_logic_1164.all

3     entity circuit_1 is

4         Port ( a : in  STD_LOGIC;

5                b : in  STD_LOGIC;

6                out1 : out  STD_LOGIC);

7     end circuit_1;

----------------------------------------------------

8     architecture Behavioral of circuit_1 is

9     begin

10          out1 <= ( a and b );

11    end Behavioral;
```

Here, we create two new lines to go above what we've already created. The first line adds the library "ieee" and the second line specifies that the package "std_logic_1164" from this library is required. Since "std_logic" is a commonly used data type, we almost always need to add the "ieee" library and the "std_logic_1164" package to the VHDL code.

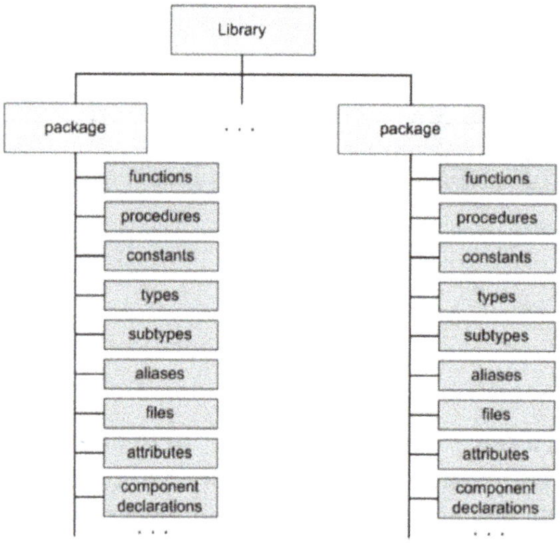

Figure: A library may consist of several packages

Now that we are familiar with the fundamental units in VHDL code, let's review one of the most important VHDL data types, i.e., the "std_logic" data type.

## The "std_logic" Data Type (vs. "bit")

As mentioned above, the "std_logic" data type can be used to represent a one-bit signal. Interestingly, there is another VHDL data type, "bit", which can take on a logic one or a logic zero.

So why do we need the std_logic data type if the "bit" data type already covers the high and low states of a digital signal? Well, a digital signal is actually not limited to logic high and logic low. Consider a tri-state inverter, as shown in figure below.

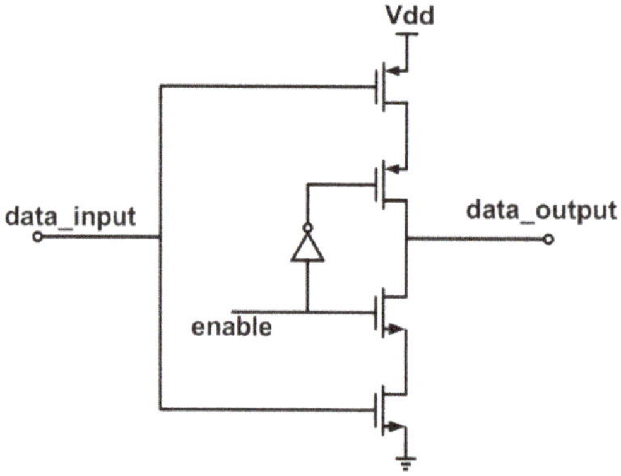

Figure: The transistor-level schematic of a tri-state inverter

When "enable" is high, "data_output" is connected to either Vdd or ground; however, when "enable" is low, "data_output" is floating, i.e., it does not have a low-impedance connection to Vdd or ground but instead presents a "high impedance" to the external circuitry. The "std_logic" data type allows us to describe a digital signal in high-impedance mode by assigning the value 'Z'.

There is another state—i.e., in addition to logic high, logic low, and high impedance—that can be used in the design of digital circuits. Sometimes we don't care about the value of a particular input.

In this case, representing the value of the signal with a "don't care" can lead to a more efficient design. The "std_logic" data type supports the "don't care" state. This enables better hardware optimization for look-up tables.

The "std_logic" data type also allows us to represent an uninitialized signal by assigning the value 'U'. This can be helpful when simulating a piece of code in VHDL. It turns out that the "std_logic" data type can actually take on nine values:

- 'U': Uninitialized

- '1' : The usual indicator for a logic high, also known as 'Forcing high'

- '0': The usual indicator for a logic low, also known as 'Forcing low'

- 'Z': High impedance

- '-': Don't care

- 'W': Weak unknown

- 'X': Forcing unknown

- 'H': Weak high

- 'L': Weak low

Among these values, we commonly use '0', '1', 'Z', and '-'.

Let's look at an example.

Example:

Write the VHDL code for the circuit in figure below.

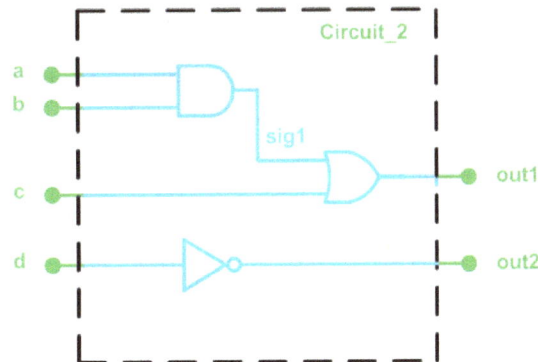

The general procedure is almost the same as the previous example. The code will be as follows:

```
1       library IEEE;

2       use IEEE.STD_LOGIC_1164.ALL;
```

------------------------------------------------------

```
3     entity circuit_2 is

4         Port ( a : in   STD_LOGIC;

5                   b : in   STD_LOGIC;

6                   c : in   STD_LOGIC;

7                   d : in   STD_LOGIC;

8                     out1 : out   STD_LOGIC;

9                     out2 : out   STD_LOGIC);

10    end circuit_2;

      -------------------------------------------------

11    architecture Behavioral of circuit_2 is

12          signal sig1: std_logic;

13    begin

14          sig1 <= ( a and b );

15          out1 <= ( sig1 or c );

16          out2 <= (not d);


17    end Behavioral;
```

Lines 1 and 2: These lines add the required library and package to the code. Since the "std_logic" data type is used, we have to add the "std_logic_1164" package.

Lines 3-10: These lines specify the name of the module along with its input/output ports. This part of the code corresponds to the parts of figure above that are in green.

Lines 11-17: This part of the code describes the operation of the circuit (those parts of figure above that are in blue). As you may have noticed, there is one internal node in figure above; it is labeled "sig1". We use the "port" statement from "entity" to define the input/output ports, but how can we define the internal nodes of a circuit? For this, we use the "signal" keyword.

In line 12 of the above code, the "signal" keyword tells the synthesis software that there is a node in the circuit labeled "sig1". Similar to the definition of the ports, we use the keyword "std_logic" after the colon to specify the required data type. Now we can assign a value to this node (line 14) or use its value (line 15).

Example:

Write the VHDL code for the circuit in figure below.

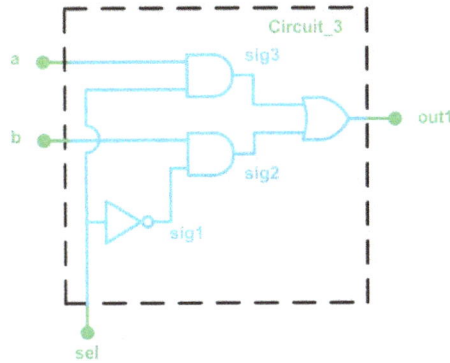

This circuit is a two-to-one multiplexer. When "sel" is high, the output of the lower AND gate will be low regardless of the value of "b". We may say that the AND gate prevents "b" from propagating to "sig2". On the other hand, since "sel" is high, the output of the upper AND gate will follow "a". Or, equivalently, "a" will reach "sig3". Since "sig2" is low in this case, the output of the OR gate will be the same as "sig3". Hence, when "sel" is high, "out1" will be the same as "a".

A similar discussion will reveal that, when "sel" is low, "out1" will take on the value of "b". Hence, based on the value of "sel", we can allow one input or the other one to reach the output. This is called multiplexing and the circuit is called a multiplexer.

We can describe the circuit of figure above using the following code:

```
1       library IEEE;

2       use IEEE.STD_LOGIC_1164.ALL;

--------------------------------------------------------

3       entity circuit_3 is

4           Port ( a : in  STD_LOGIC;

5                    b : in  STD_LOGIC;

6                    sel : in  STD_LOGIC;

7                    out1 : out  STD_LOGIC);

8       end circuit_3;

--------------------------------------------------------

9       architecture Behavioral of circuit_3 is

10          signal sig1, sig2, sig3: std_logic;

11      begin

12          sig1 <= ( not sel );

13          sig2 <= ( b and sig1 );

14          sig3 <= ( a and sel );
```

```
15          out1 <= ( sig2 or sig3 );


16     end Behavioral;
```

## Register-transfer Level

Register Transfer Level (RTL) is an abstraction for defining the digital portions of a design. It is the principle abstraction used for defining electronic systems today and often serves as the golden model in the design and verification flow. The RTL design is usually captured using a hardware description language (HDL) such as Verilog or VHDL. While these languages are capable of defining systems at other levels of abstraction, it is generally the RTL semantics of these languages, and indeed a subset of these languages defined as the synthesizable subset. This means the language constructs that can be reliably fed into a logic synthesis tool which in turn creates the gate-level abstraction of the design that is used for all downstream implementation operations.

RTL is based on synchronous logic and contains three primary pieces namely, registers which hold state information, combinatorial logic which defines the nest state inputs and clocks that control when the state changes.

### RTL Design

Register Transfer Level, or RTL1 design lies between a purely behavioral description of the desired circuit and a purely structural one. An RTL description de- scribes a circuit's registers and the sequence of transfers between these registers but does not describe the hardware used to carry out these operations.

As a simple example, consider a device that needs to add four numbers. In VHDL, given signals of the correct type, we can simply write:

s <= a + b + c + d ;

This particular description is simple enough that it can be synthesized. However, the resulting circuit will be a fairly large combinational circuit comprising three adder circuits as follows:

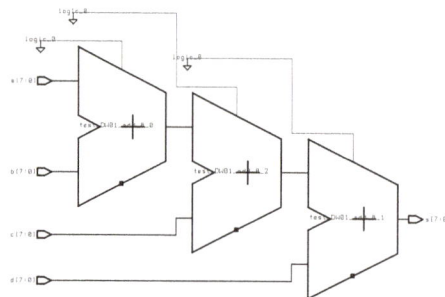

A behavioral description, not being concerned with implementation details would be complete at this point.

However, if we were concerned about the cost of the implementation we might decide to break down the computation into a sequence of steps, each one involving only a single addition:

s = 0

s = s + a

s = s + b

s = s + c

s = s + d

where each operation is executed sequentially. The logic required is now one adder, a register to hold the value of s in-between operations, a multiplexer to select the input to be added on, and a circuit to clear s at the start of the computation.

Although this approach only needs one adder, the process requires more steps and will take longer. Circuits that divide up a computation into a sequence of arithmetic and logic operations are quite common and this type of design is called Register Transfer Level (RTL) or "dataflow" design.

An RTL design is composed of (1) registers and combinational function blocks (e.g. adders and multiplexers) called the *data path* and (2) a finite state machine, called the *controller* that controls the transfer of data through the function blocks and between the registers.

In VHDL RTL design the gate-level design and optimization of the datapath (registers, multiplexers, and combinational functions) is done by the synthesizer. However, the designer must design the state machine and decide which register transfers are performed in which state.

The RTL designer can trade off datapath complexity (e.g. using more adders and thus using more chip area) against speed (e.g. having more adders means fewer steps are required to obtain the result). RTL design is well suited for the design of micro- processors and special-purpose processors such as disk drive controllers, video display cards, network adapter cards, etc. It gives the designer great flexibility in choosing between processing speed and circuit complexity.

The diagram below shows a generic component in the datapath. Each RTL design will be composed of one of the following building blocks for each register. The structure allows the contents of each register to be updated at the end of each clock period with a value selected by the controller. The widths of the registers, the types of combinational functions and their inputs will be determined by the application. A typical design will include many of these components.

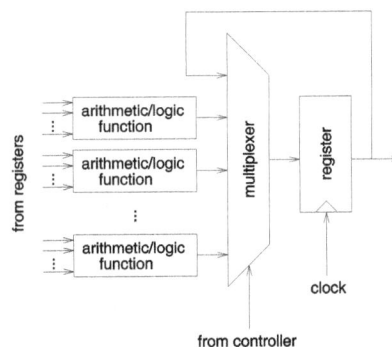

## RTL Design Example

To show how an RTL design is described in VHDL and to clarify the concepts involved, we will design a four-input adder. This design will also demonstrate how to create packages of components that can be re-used.

The datapath shown below can load the register at the start of each clock cycle with zero, the current value of the register, or the sum of the register and one of the four inputs. It includes one 8-bit register, an 8-bit adder and a multiplexer that selects one of the four inputs as the value to be added to the current value of the register.

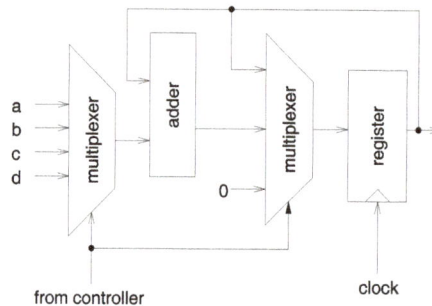

The first design unit is a package that defines a new type, num, for eight-bit unsigned numbers and an enumerated type, states, with six possible values. nums are defined as a subtype of the unsigned type

```
-- RTL design of 4-input summer

-- subtype used in design

library ieee ;

use ieee.std_logic_1164.all ;

use ieee.std_logic_arith.all ;

package averager_types is

subtype num is unsigned (7 downto 0) ;

type states is (clr, add_a, add_b, add_c,

add_d, hold) ;

end averager_types ;
```

The first entity defines the datapath. In this case the four numbers to be added are available as inputs to the entity and there is one output for the current sum.

The inputs to the datapath from the controller are a 2-bit selector for the multiplexer and two control signals to load or clear (set to 0) the register.

```vhdl
-- datapath
library ieee ;
use ieee.std_logic_1164.all ;
use ieee.std_logic_arith.all ;
use work.averager_types.all ;
entity datapath is
port (
a, b, c, d : in num ;
sum : out num ;
sel : in std_logic_vector (1 downto 0) ;
load, clear, clk : in std_logic
) ;
end datapath ;
architecture rtl of datapath is
signal mux_out, sum_reg, next_sum_reg : num ;
constant sum_zero : num :=
conv_unsigned(0,next_sum_reg'length) ;
begin
-- mux to select input to add
with sel select mux_out <=
a when "00",
b when "01",
c when "10",
d when others ;
-- mux to select register input
next_sum_reg <=
sum_reg + mux_out when load = '1' else
sum_zero when clear = '1' else
```

```
sum_reg ;

-- register sum

process(clk)

begin

if clk'event and clk = '1' then

sum_reg <= next_sum_reg ;

end if ;

end process ;

-- entity output is register output

sum <= sum_reg ;

end rtl ;
```

The RTL design's controller is a state machine whose outputs control the multiplexers in the data-path. The controller's inputs are signals that control the controller's state transitions. In this case the only input is an update signal that tells our device to re- compute the sum (presumably because one or more of the inputs has changed).

This particular state machine sits at the "hold" state until the update signal is true. It then sequences through the other five states and then stops at the hold state again. The other five states are used to clear the register and to add the four inputs to the current value of the register.

```
-- controller
library ieee ;
use ieee.std_logic_1164.all ;
use work.averager_types.all ;
entity controller is
port (
update : in std_logic ;
sel : out std_logic_vector (1 downto 0) ;
load, clear : out std_logic ;
clk : in std_logic
) ;
```

```vhdl
end controller ;
architecture rtl of controller is
signal s, holdns, ns : states ;
signal tmp : std_logic_vector (3 downto 0) ;
begin
-- select next state
with s select ns <=
add_a when clr,
add_b when add_a,
add_c when add_b,
add_d when add_c,
hold when add_d,
holdns when others ; -- hold
-- next state if in hold state
holdns <=
clr when update = '1' else
hold ;
-- state register
process(ns,clk)
begin
if clk'event and clk = '1' then
s <= ns ;
end if ;
end process ;
-- controller outputs
with s select sel <=
"00" when add_a,
"01" when add_b,
```

```
"10" when add_c,

"11" when others ;

load <= '0' when s = clr or s = hold else '1' ;

clear <= '1' when s = clr else '0' ;

end rtl ;
```

The code is an example of how the datapath and the controller entities can be placed in a package, average_components, as components. In practice the datapath and controller component declarations would probably have been placed in the top-level architecture since they are not likely to be re-used in other designs.

```
-- package for datapath and controller

library ieee ;

use ieee.std_logic_1164.all ;

use work.averager_types.all ;

package averager_components is

component datapath

port (

a, b, c, d : in num ;

sum : out num ;

sel : in std_logic_vector (1 downto 0) ;

load, clear, clk : in std_logic

) ;

end component ;

component controller

port (

update : in std_logic ;

sel : out std_logic_vector (1 downto 0) ;

load, clear : out std_logic ;

clk : in std_logic

) ;
```

```
            end component ;

            end averager_components ;
```

The top-level averager entity instantiates the two components and interconnects them.

```
            -- averager

            library ieee ;

            use ieee.std_logic_1164.all ;

            use ieee.std_logic_arith.all ;

            use work.averager_types.all ;

            use work.averager_components.all ;

            entity averager is port (

            a, b, c, d : in num ;

            sum : out num ;

            update, clk : in std_logic ) ;

            end averager ;

            architecture rtl of averager is

            signal sel : std_logic_vector (1 downto 0) ;

            signal load, clear : std_logic ;

            -- other declarations (e.g. components) here

            begin

            d1: datapath port map ( a, b, c, d, sum, sel, load,

            clear, clk ) ;

            c1: controller port map ( update, sel, load,

            clear, clk ) ;

            end rtl ;
```

The result of the synthesizing the datapath is:

The register flip-flops are at the upper right, the adder is in the middle and the input multiplexer is at the lower left.

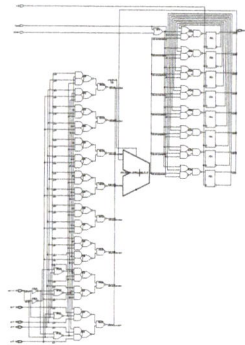

The result of the synthesizing the controller is:

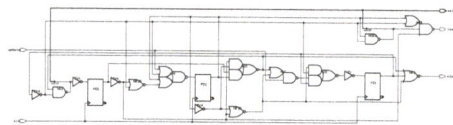

The following timing diagram shows the datapath output and the controller state over one computation. Note that the state and output transitions take place on the rising edge of the clock. Also note that the output is updated at the end of the state in which a particular operation is performed.

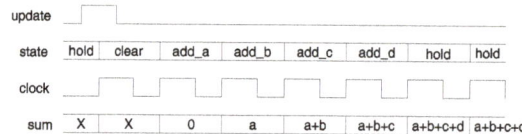

| update | | | | | | | | | |
|--------|------|-------|-------|-------|-------|-------|-----------|-----------|
| state  | hold | clear | add_a | add_b | add_c | add_d | hold | hold |
| clock  | | | | | | | | | |
| sum    | X | X | 0 | a | a+b | a+b+c | a+b+c+d | a+b+c+d |

## RTL Timing Analysis

The datapath is a synchronous sequential circuit that uses the same clock for all registers and all register contents thus change at the same time. The controller uses the same clock as the datapath. Each datapath register loads the values "computed" during one state at the end of that state (which is also the start of the next state).

We can guarantee that the correct results will be loaded into registers if the worst-case propagation delay ($t_{PD}$) through any path of multiplexers and combinational function blocks is less than the clock period ($t_{clock}$) minus the registers' setup time ($t_s$) and clock-to-output ($t_{CO}$) delays

$$t_{PD} < t_{clock} - t_s - t_{CO}$$

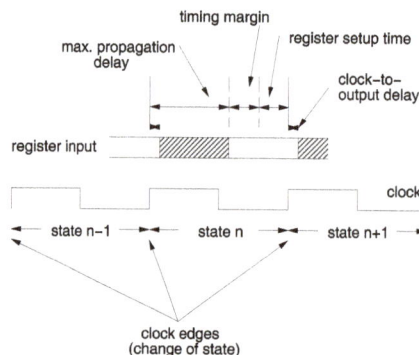

Using a single clock means we only need to compute the delay through combinational logic blocks which is much simpler than having to deal with asynchronous clocks. This is why almost all large-scale digital circuits are synchronous designs.

Synthesis tools can be asked to synthesize logic that operates at a particular clock period. The synthesizer is supplied with the propagation delay specifications for the combinational logic components available in the particular technology being used and it will then try to arrange the logic so that the propagation delay from any input or register output to the inputs of all registers is less than the clock period. This ensures that the circuit will work properly at the specified clock rate

## Behavioral Synthesis

It is possible to work at even higher levels of abstraction than RTL when design time is more important than cost. Advanced synthesis programs (for example, Synopsys' Behavioral Compiler) can convert a behavioral description of an algorithm into an RTL description. The compiler does this by automatically allocating registers and partitioning the processing over as many clock cycles as are required to meet high-level processing time requirements.

## Register Transfer Language

Register Transfer Language, RTL, (sometimes called register transfer notation) is a powerful high level method of describing the architecture of a circuit. VHDL code and schematics are often created from RTL. RTL describes the transfer of data from register to register, known as microinstructions or micro operations. Transfers may be conditional. Each microinstruction completes in one clock cycle. A typical RTL statement will look like the following:

$A \vee B \; --> \; R1 \; <-- \; R2;$

This is read as "if signal A or signal B is true then register R2 is transferred to register R1". The first part, A v B, is a logical expression that must be true for the transfer to take place. The --> symbol separates the logical expression from the microinstruction. It is the if-then part of the statement. If there isn't a logical expression, --> isn't in the statement and the microinstruction will always take place. To the right of --> is the microinstruction. It describes a transfer of data and operations on the data from register to register. The above RTL statement is equivalent to the following schematic:

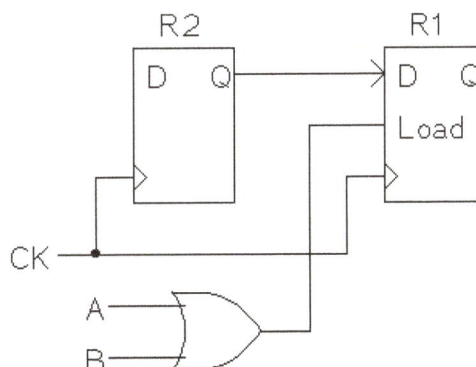

For RTL we will use the following symbols:

| | |
|---|---|
| <-- | Register transfer |
| [ ] | Word index |
| < > | Bit index |
| n..m | Index range |
| --> | If-then |
| := | Definition |
| # | Concatenation |
| : | Parallel separator |
| ; | Sequential separator |
| @ | Replication |
| { } | Operation modifier |
| ( ) | Operation or value grouping |
| = != < <= > >= | Comparison operators |
| + - * / | Arithmetic operators |
| ^ v ' xor | Logical operators |

## RTL Examples

Example 1:

$K1 \; --> \; R0 \; <-- \; R1 : K1' \; \wedge \; K2 \; --> \; R0 \; <-- \; R2 ;$

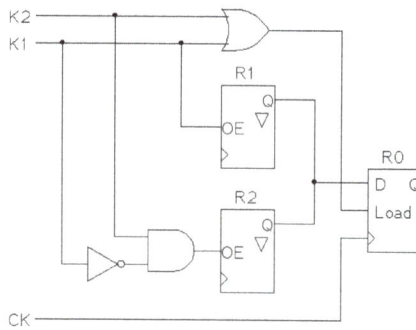

Example 2:

$R0 \; <-- \; R1 + R2' + 1$

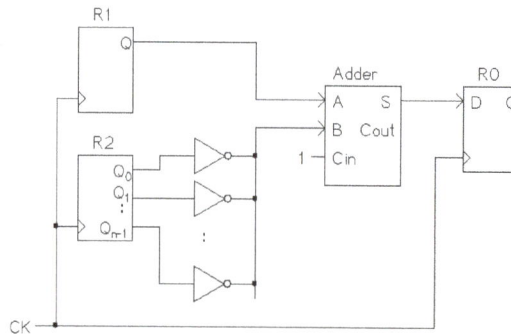

The above circuit is equivalent to *R0 <-- R1 - R2*

We will learn RTL with a very simple processor called a "special math processor".

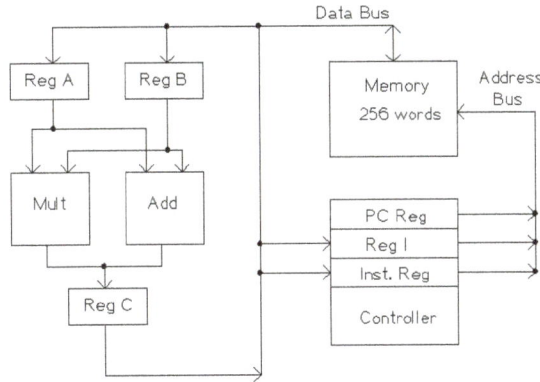

## Processor State (Using RTL)

| | |
|---|---|
| RA<15..0>: | Register A (input to multiplier and adder) |
| RB<15..0>: | Register B (input to multiplier and adder) |
| RC<15..0>: | Register C (output from multiplier or adder) |
| PC<7..0>: | Program Counter (Address of next instruction) |
| RI<7..0>: | Register I (memory index register) |
| IR<11..0>: | Instruction Register |
| Reset: | Reset signal |
| op<3..0> := IR<11..8>: | Operation code field |
| M[255..0]<15..0>: | Main memory 255 words. |

## Processor Schematic

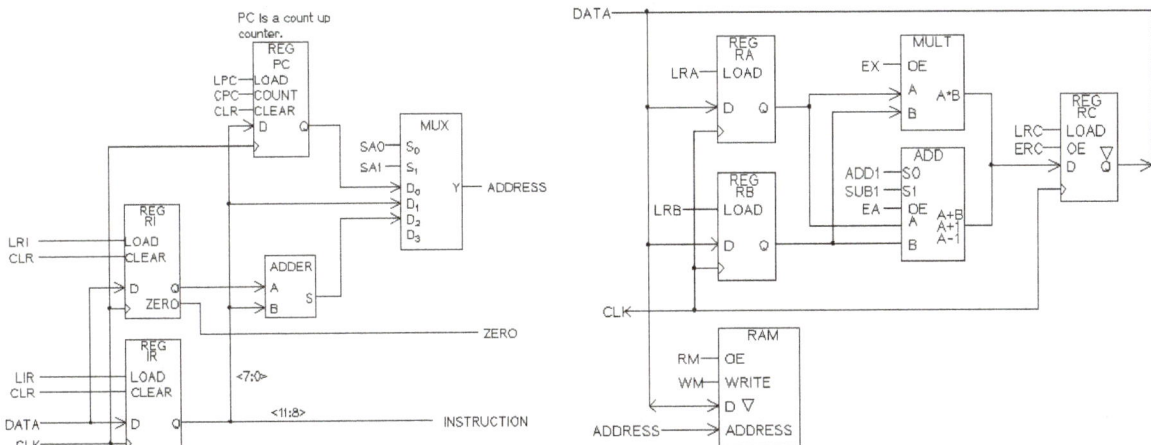

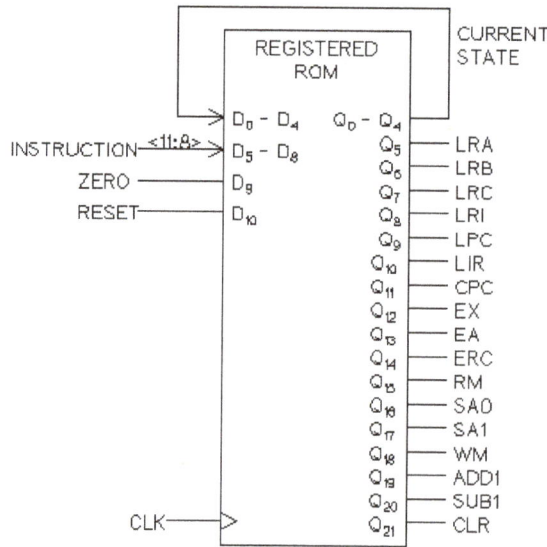

## Processor Instructions

- Instructions 0 through 3 are memory/register transfers
- Instructions 4 is a register to register transfer
- Instructions 5 through 8 are math instructions
- Instruction 9 through B are processor control instruction

| Code | Instruction Syntax | Operation |
|------|-------------------|-----------|
| 0x00XX | load RA=[XX+RI] | Load RA from M[XX+RI<7..0>] |
| 0x01XX | load RB=[XX+RI] | Load RB from M[XX+RI<7..0>] |
| 0x02XX | load RI=[XX] | Load RI from M[XX] |
| 0x03XX | load [XX+RI]=RC | Load M[XX+RI<7..0>] from RC |
| 0x04XX | mov RB=RC | Move RC to RB |
| 0x05XX | add | Add RA and RB and put in RC |
| 0x06XX | mult | Multiply RA and RB and put in RC |
| 0x07XX | inc | Increment RA and put in RC |
| 0x08XX | dec | Decrement RA and put in RC |
| 0x09XX | jmp XX | Jump to XX |
| 0x0AXX | jz XX | Jump to XX if RI is zero |
| 0x0BXX | halt | Halt |

## Fetch Execute Cycle

```
instruction_interpretation := (Reset --> (PC <-- 0: IR <-- 0: RI <-- 0: RA
<-- 0:
```

                                                        RB <-- 0: RC <-- 0);
instruction_interpretation):

                              Reset' --> (IR <-- M[PC<7..0>]<11..0>: PC <-
-PC + 1; instruction_execution))


instruction_execution := ( (ldra (:= op = 0) --> RA <--
M[(IR<7..0>+RI<7..0>)^0xff]:

                         Continue with remaining opcodes:

                   halt (:= op = 11) --> instruction_execution);

                   instruction_interpretation)


## Possible RTL Microinstructions

PC <-- 0x00

PC <-- IR<7..0>

PC <-- PC + 1

RA <-- M[(IR<7..0>+RI<7..0>)^0xff]

RB <-- M[(IR<7..0>+RI<7..0>)^0xff]

RB <-- RC

RC <-- RA - 1

RC <-- RA + 1

RC <-- RA * RB

RC <-- RA + RB

RI <-- M[IR<7..0>]<7..0>

RI <-- 0x00

IR <-- M[PC<11..0>]

IR <-- 0x00

M[(IR<7..0>+RI<7..0>)^0xff] <-- RC

# Pipeline

Pipelining is the process of accumulating instruction from the processor through a pipeline. It allows storing and executing instructions in an orderly process. It is also known as pipeline processing.

Pipelining is a technique where multiple instructions are overlapped during execution. Pipeline is divided into stages and these stages are connected with one another to form a pipe like structure. Instructions enter from one end and exit from another end.

Pipelining increases the overall instruction throughput.

In pipeline system, each segment consists of an input register followed by a combinational circuit. The register is used to hold data and combinational circuit performs operations on it. The output of combinational circuit is applied to the input register of the next segment.

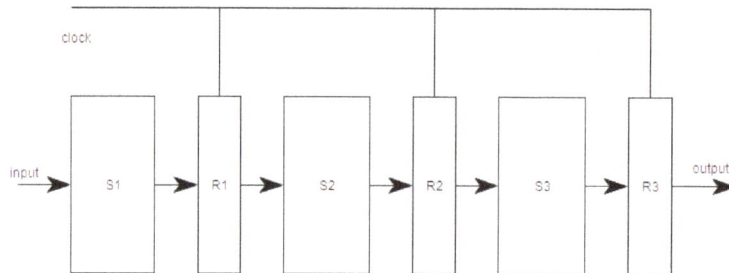

Pipeline system is like the modern day assembly line setup in factories. For example in a car manufacturing industry, huge assembly lines are setup and at each point, there are robotic arms to perform a certain task, and then the car moves on ahead to the next arm.

## Types of Pipeline

It is divided into 2 categories:

1. Arithmetic Pipeline

2. Instruction Pipeline

## Arithmetic Pipeline

Arithmetic pipelines are usually found in most of the computers. They are used for floating point operations, multiplication of fixed point numbers etc. For example: The input to the Floating Point Adder pipeline is:

$X = A*2^{\wedge}a$

$Y = B*2^{\wedge}b$

Here A and B are mantissas (significant digit of floating point numbers), while a and b are exponents.

The floating point addition and subtraction is done in 4 parts:

1.  Compare the exponents.

2.  Align the mantissas.

3.  Add or subtract mantissas.

4.  Produce the result.

Registers are used for storing the intermediate results between the above operations.

## Instruction Pipeline

In this a stream of instructions can be executed by overlapping *fetch*, *decode* and *execute* phases of an instruction cycle. This type of technique is used to increase the throughput of the computer system.

An instruction pipeline reads instruction from the memory while previous instructions are being executed in other segments of the pipeline. Thus we can execute multiple instructions simultaneously. The pipeline will be more efficient if the instruction cycle is divided into segments of equal duration.

## Pipeline Conflicts

There are some factors that cause the pipeline to deviate its normal performance. Some of these factors are given below:

## Timing Variations

All stages cannot take same amount of time. This problem generally occurs in instruction processing where different instructions have different operand requirements and thus different processing time.

## Data Hazards

When several instructions are in partial execution, and if they reference same data then the problem arises. We must ensure that next instruction does not attempt to access data before the current instruction, because this will lead to incorrect results.

## Branching

In order to fetch and execute the next instruction, we must know what that instruction is. If the present instruction is a conditional branch, and its result will lead us to the next instruction, then the next instruction may not be known until the current one is processed.

## Interrupts

Interrupts set unwanted instruction into the instruction stream. Interrupts effect the execution of instruction.

### Data Dependency

It arises when an instruction depends upon the result of a previous instruction but this result is not yet available.

### Advantages of Pipelining

1. The cycle time of the processor is reduced

2. It increases the throughput of the system

3. It makes the system reliable.

### Disadvantages of Pipelining

1. The design of pipelined processor is complex and costly to manufacture

2. The instruction latency is more.

## Globally Asynchronous Locally Synchronous

Clock generation and distribution on microprocessors is becoming more challenging with advances in VLSI technology. Higher levels of integration and deeper pipelines on larger dies increase the total clock load. More clock buffers are required, increasing the clock distribution latency. These buffers introduce more skew due to reduced manufacturing control of shrinking geometries. Higher frequencies give rise to more power supply fluctuations and more cross-coupling, and this noise increases clock jitter. All of these effects lead to more clock power. Furthermore, as skew and jitter become a higher percentage of shorter clock cycle times, the fraction of time available for logic evaluation is reduced.

Many recent microprocessor designs address the clock design challenge by relaxing the amount of skew which is tolerable in early stages of the clock distribution and compensating for the variation at each lower-level clock domain. The Alpha processor strives for zero skew with DLL-based phase locking, while the Pentium 4 processor uses a programmed nonzero inter-domain skew. In both cases, traditional timing analysis is used to verify that setup and hold time requirements on inter domain signals are satisfied.

These methodologies are similar to the globally asynchronous locally-synchronous (GALS) design style, in which the system is partitioned into synchronous blocks (SBs) of logic which communicate with each other asynchronously. The key difference is that true GALS architectures allow arbitrary skew between clock domains and use some form of synchronization for inter-block communication. Unfortunately, these synchronization strategies are often a source of nondeterminism, which greatly complicates validation, debug, and test.

### Nondeterminism in GALS Systems

A system is nondeterministic if there are multiple possible sequences of states and outputs with which it may correctly respond to a given input sequence. Nondeterminism is not necessarily indicative of a faulty design, since an implementation is considered correct as long as it conforms to its higher-level specification.

Synchronous systems are typically designed to be deterministic. The next state and outputs are uniquely determined by the current state and inputs. All signals which are sampled by clocks are designed through worstcase timing analysis to be stable at their final logic value long enough before the clock edge that the sampling state element's setup time is satisfied.

Similarly, sampling the states and outputs of a SB in a GALS system with its local clock will produce deterministic state and output sequences in response to a given input sequence to the SB. In most GALS methodologies, however, asynchronous inputs to SBs are captured by synchronizers, so that the relative order of input transitions and clock transitions is unpredictable. This makes the input sequence, and therefore the state and output sequences, of the SB nondeterministic. For one SB input signal and one clock edge, there are two possible next states. In a GALS system with hundreds of asynchronous bits switching for thousands of clock cycles, the number of possible state sequences combinatorially explodes. As a result, the system output sequence may differ when the input sequence is applied to multiple copies of the same design or when the input sequence is repeatedly applied to a single instance of the design.

Metastability, the condition where the output of a synchronizer is neither 0 nor 1 for a period of time, is a special case of nondeterminism which occurs when the time separation of the signal and clock transitions is very small . While metastability is also undesirable, the lack of it does not imply deterministic behavior.

A deterministic GALS system must handle the synchronization of inter-SB signals such that the input sequence presented to the SB is unique despite variation in clock skew, clock frequencies, and interconnect delays. However, since the skew between clocks in different SBs is uncontrolled, the total state of a deterministic GALS system at any instant in time is not unique, even though the sequences in each SB are unique.

## Nondeterminism in a Processor

Nondeterminism can be observed in a GALS implementation of an out-of-order processor core which was implemented in Verilog, an environment which is able to simulate concurrency and nonzero delays. The processor core consists of a register file and four ALUs. Each ALU can perform the functions add, subtract, multiply, and move (copy). The register file has two read ports which are used simultaneously by a single ALU to read its operands. An arbiter assigns a static priority to each ALU and grants access to the register file's read ports to the ALU with the highest priority request. The register file also has one write port, managed by a separate arbiter, through which an ALU writes its result. Out-of-order execution is supported with a scoreboard, which controls four of the stages of an instruction's life cycle: issue, read operands, execute, and write result. The system consists of five synchronous blocks: one for each ALU and one for the register file and scoreboard. While this partitioning may not be practical in terms of area or performance, it allows nondeterminism to be seen easily at the behavioral level. All five clocks in the system run at the same frequency, although this is not generally true of GALS systems.

Consider the following in-order instruction sequence. Instructions are named I1 - I5. Registers are named R1 - R7. The destination register is always the first argument in the list, followed by one or two source registers.

I1: ADD R3, R1, R2

I2: MUL R5, R3, R4

I3: SUB R4, R2, R1

I4: MOV R6, R3

I5: ADD R4, R3, R2

The architectural spec for the processor defines a partial order of register read and write events to ensure the avoidance data hazards. These include RAW hazards between I1 and {I2, I4, I5}, a WAR hazard between I2 and I3; and a WAW hazard between I3 and I5. The architectural spec does not impose any constraints on the relative order of independent events, such as accesses to different registers.

Tables show three possible traces of the execution of the above instructions generated by varying clock phases and handshake wire delays. Each column corresponds to an instruction, and each row corresponds to a cycle of the register file/scoreboard SB clock. Table entries indicate the clock cycle on which each instruction stage completes.

Table is used as a baseline against which other traces may be compared.

Table shows the effect of increased delay on the asynchronous handshake wire which is asserted by the ALU executing instruction I3 to indicate to the scoreboard that the result is ready. I3's execution and write stages are postponed by 1 clock cycle, and the WAW hazard between I3 and I5 postpones I5's entire execution sequence by 1 clock cycle. The cycles during which I2 and I4 write are unchanged. Likewise, the relative order in which all instructions write is unchanged

Table shows the effect of changing the clock phase of the ALU executing instruction I4. Because less time is spent on the synchronization of the handshakes, I4 finishes execution 1 scoreboard-clock cycle early. Since I4's ALU is no longer competing with I2's higher priority ALU for access to the write bus, the arbiter allows I4's write to occur before I2's, changing the sequence of writes.

The final state of the register file following the execution of all instructions is identical in all three scenarios. The out-of-order processor thus conforms to the architectural spec by correctly executing the instruction sequences, even though its intermediate sequences and cycle-by-cycle behavior vary due to clock skew and wire delays.

| Cycle | I1 | I2 | I3 | I4 | I5 |
|-------|-------|-------|-------|-------|-----|
| 1 | Issue | | | | |
| 2 | Read | Issue | | | |
| 3 | | | Issue | | |
| 4 | | | Read | Issue | |
| 5 | Exec | | | | |

| 6 | Write | | | |
|---|---|---|---|---|
| 7 | | Read | Exec | |
| 8 | | | Write | Read |
| 9 | | | | | Issue |
| 10 | | | | Exec | Read |
| 11 | | Exec | | | |
| 12 | | Write | | | |
| 13 | | | | Write | Exec |
| 14 | | | | | Write |

Table: Baseline trace of instruction execution.

| Cycle | I1 | I2 | I3 | I4 | I5 |
|---|---|---|---|---|---|
| 1 | Issue | | | | |
| 2 | Read | Issue | | | |
| 3 | | | Issue | | |
| 4 | | | Read | Issue | |
| 5 | Exec | | | | |
| 6 | Write | | | | |
| 7 | | Read | | | |
| 8 | | | Exec | Read | |
| 9 | | | Write | | |
| 10 | | | | | Issue |
| 11 | | Exec | | Exec | Read |
| 12 | | Write | | | |
| 13 | | | | Write | |
| 14 | | | | | Exec |
| 15 | | | | | Write |

Table: Slower handshakes with I3's ALU.

| Cycle | I1 | I2 | I3 | I4 | I5 |
|---|---|---|---|---|---|
| 1 | Issue | | | | |
| 2 | Read | Issue | | | |
| 3 | | | Issue | | |
| 4 | | | Read | Issue | |
| 5 | Exec | | | | |
| 6 | Write | | | | |
| 7 | | Read | Exec | | |
| 8 | | | Write | Read | |
| 9 | | | | | Issue |
| 10 | | | | Exec | Read |
| 11 | | Exec | | Write | |
| 12 | | Write | | | |
| 13 | | | | | Exec |
| 14 | | | | | Write |

Table: Clock phase difference in I4's ALU.

## Impact on Validation, Debug and Test

Nondeterminism makes simulation-based validation more expensive. The simulator must choose among multiple possible next state and output values. Simulation must be repeated for many different choices to ensure that the design conforms to the spec regardless of the nondeterministic outcome. Trying to avoid this cost by validating only individual SBs risks missing bugs associated with complex system-level behaviors.

Nondeterminism thwarts the use of test techniques which perform cycle-by-cycle comparisons of observed and expected response sequences, such as the clock gating validation in . Comparing Table with Table, for example, results in a mismatch in the state of I3's destination register in cycle 8. If the test response analyzer adapts to the difference by postponing the entire expectation by one clock cycle, the writes of I2 and I4, which occur on schedule, will cause mismatches. If no adjustment is made, the write of I5 will cause another mismatch. In either case, it may not be clear whether the mismatch is the result of excessive delay on an asynchronous signal (which is acceptable) or a critical path violation within a SB (which could cause an unacceptable deviation from the spec for some other instruction sequence).

Observing the system only after the test reaches a deterministic point, e.g. after all active instructions have completed, may provide insufficient observability. Only architectural state, such as the contents of the register file, would be eligible for observation since other internal state is not included in the spec. Observation points may be few and temporally distant, making root-cause analysis very difficult.

Event-based testers can handle a limited amount of nondeterminism by processing signal transitions on pins which need not be mapped to specific clock cycles. However, this approach is not effective for scan tests which shift out internal state captured on a specific clock cycle. It is also inapplicable to tests in which the sequence of events is changed as in table, where the state of the register file after cycle 11 is one which is never reached by the expectation in table.

Nondeterminism precludes the use of many powerful silicon debug techniques. Waveform acquisition using optical probing relies on a deterministic response to lock onto the trigger transition each time through the loop. Shmoo plotting uses output sequence mismatches to identify the boundaries of acceptable operating regions. Much of the debug of the McKinley processor is performed using a tester which stays synchronized with the internal state of the chip rather than a system platform in which asynchronous system events such as memory refreshes and interrupts cause nondeterministic behavior.

There is a high simulation and test application time cost associated with generating all possible correct responses for each test pattern. Storage of those responses on-chip for BIST costs precious die area, while off-chip storage requires either a large, expensive, high speed memory or a further increase in test time for repeated memory reloads . If real faults map to an alternative valid output sequence, there may be a decrease in fault coverage and an increase in escape rate.

A variety of GALS architectures have been proposed which use synchronizers, stoppable clocks, or both to sample asynchronous signals with a clock.

SBs whose inputs are sampled by flip-flops with internal metastability detectors and which stop the local clock until the metastability resolves itself have been proposed. Since the metastability can persist for an unbounded amount of time, some schemes don't update a state which remains metastable after a certain period of time . Some methodologies only synchronize data request lines and ensure through careful design that bundled data is valid before the request is asserted. They arbitrate between incoming requests and the local clock in a variety of ways: using the clock as a non-persistent arbiter input , generating a clock disable signal , or inserting an arbiter directly into the ring oscillator  . However, all of these cases are nondeterministic because whether an input transition occurs before or after a particular local clock edge is unpredictable.

GALS architectures which achieve deterministic behavior do so by imposing constraints on their environments. Some require that incoming requests occur with such low frequency that all local processing completes and the local clock is stopped before the next request arrives . Others require that both SBs receive different fanout branches of the same global clock. Data is added to and removed from the FIFO at the same rate so that the FIFO never becomes empty or full . These methodologies are not applicable to blocks with high I/O bandwidths and sporadic workloads, such as a specialized execution unit in a microprocessor.

Chapiro  described an escapement organization which uses handshaking signals following a known protocol to restart a stopped clock. Unlike data signals which may or may not transition before they are ready to be sampled again, all of the information in the handshaking signal is contained in its transition time rather than its logic level. Consequently, the asynchronous signal does not require synchronization and thus poses no risk of metastability.

A token ring communication protocol known as FDDI  includes counters at each node which keep track of the length of time the token has been held and the length of time since the token was last released.

## Synchronous-tokens System Architecture

As shown in figure, a synchronous-tokens system consists of a collection of SBs surrounded by wrapper logic and connected with asynchronous communication channels and token rings. The wrapper logic, shown in figure, consists of token ring nodes, asynchronous interfaces, and a stoppable clock. One or more SBs are designated as I/O SBs and are synchronized to and communicate with the environment (a board or a tester) without any intervening wrapper logic.

The asynchronous communication channels transport data between pairs of synchronous blocks. These channels are optionally pipelined with self-timed FIFOs to improve performance and to avoid the need for wave pipelining . At each end of a channel is an asynchronous interface; this piece of wrapper logic converts between synchronous valid bits and pulsed request and acknowledge handshakes. Data is transmitted using a bundled data signaling convention .

Each pair of communicating SBs has a token ring with a node in each SB's wrapper logic. A single token ring regulates the operation of all asynchronous communication channels in both directions between the two SBs. An asynchronous interface is enabled only while the associated token ring node in its SB is holding the token. The communication channel is designed such that the propagation delay of the token is no faster than that of the data. The effectively synchronizes the data to the clock of SB whose interface is enabled. It prevents a transmitting SB from adding data to an empty channel and producing a request which reaches the receiver on a nondeterministic cycle of its local clock. It also prevents a receiving SB from removing data from a full channel and producing an acknowledge which reaches the transmitter on a nondeterministic cycle of its local clock. An n-stage self-timed FIFO in the channel allows up to n words of data to be exchanged per token cycle.

One of the two connections which form a token ring must be inverting so that there is an odd number of inversions around the ring; this characteristic of all handshaking loops is needed for sustained oscillation. In the synchro-tokens methodology, the token rings use transition signaling in a two-phase handshaking protocol.

The node is a synchronous state machine clocked by the SB's stoppable clock, the frequency of which can be digitally controlled with either variable delay inverters or a clock divider circuit. The node connects to the token ring through the TokenIn and TokenOut signals. The node produces a FIFO clock enable, Fclken, for its associated channels which gates a branch of the stoppable clock before it reaches the asynchronous interface. The node also generates an enable for the stoppable clock itself, SBclken; the enables from all nodes in the SB are AND together so that the clock stops when any node de-asserts its SBclken.

Each node contains a pair of decrementing counters, each of which is parallel loadable from a dedicated register, which may in turn be downloadable from an on die fuse array. The "hold counter" and register control how long the node holds the token before passing it to the other node on the token ring. The "recycle counter" and register control how long after passing the token to the other node it expects to get the token back.

When the incoming token has arrived and the recycle counter reaches zero, the interfaces of the node's associated asynchronous interfaces are enabled. The hold counter decrements by one for each local clock cycle. When the hold counter reaches zero, the token is passed and the channel interfaces are disabled. The recycle counter then decrements by one for each local clock cycle. During this recycle time, local processing continues and access is granted to any other communication channel whose associated node is holding its ring's token. If the token has not returned by the time the recycle counter reaches zero, the clock to the entire SB is synchronously stopped. When the late token eventually returns, the clock is asynchronously restarted. This figure ensures that SB input data is received on a deterministic local clock cycle.

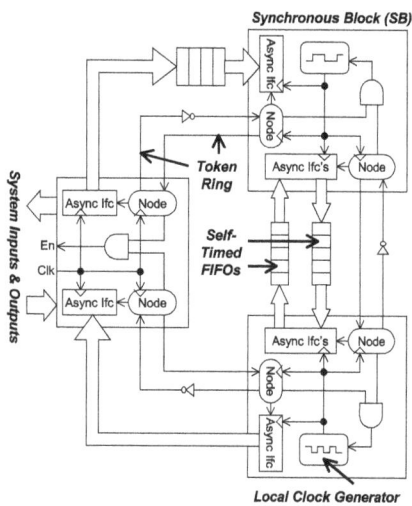

Figure: Synchro-tokens system architecture.

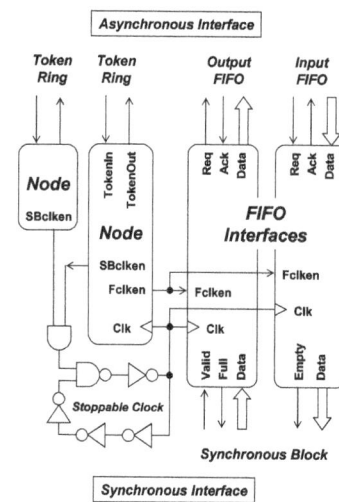

Figure: Wrapper logic, Nodes, FIFO interfaces, and a stoppable clock.

# Permissions

All chapters in this book are published with permission under the Creative Commons Attribution Share Alike License or equivalent. Every chapter published in this book has been scrutinized by our experts. Their significance has been extensively debated. The topics covered herein carry significant information for a comprehensive understanding. They may even be implemented as practical applications or may be referred to as a beginning point for further studies.

We would like to thank the editorial team for lending their expertise to make the book truly unique. They have played a crucial role in the development of this book. Without their invaluable contributions this book wouldn't have been possible. They have made vital efforts to compile up to date information on the varied aspects of this subject to make this book a valuable addition to the collection of many professionals and students.

This book was conceptualized with the vision of imparting up-to-date and integrated information in this field. To ensure the same, a matchless editorial board was set up. Every individual on the board went through rigorous rounds of assessment to prove their worth. After which they invested a large part of their time researching and compiling the most relevant data for our readers.

The editorial board has been involved in producing this book since its inception. They have spent rigorous hours researching and exploring the diverse topics which have resulted in the successful publishing of this book. They have passed on their knowledge of decades through this book. To expedite this challenging task, the publisher supported the team at every step. A small team of assistant editors was also appointed to further simplify the editing procedure and attain best results for the readers.

Apart from the editorial board, the designing team has also invested a significant amount of their time in understanding the subject and creating the most relevant covers. They scrutinized every image to scout for the most suitable representation of the subject and create an appropriate cover for the book.

The publishing team has been an ardent support to the editorial, designing and production team. Their endless efforts to recruit the best for this project, has resulted in the accomplishment of this book. They are a veteran in the field of academics and their pool of knowledge is as vast as their experience in printing. Their expertise and guidance has proved useful at every step. Their uncompromising quality standards have made this book an exceptional effort. Their encouragement from time to time has been an inspiration for everyone.

The publisher and the editorial board hope that this book will prove to be a valuable piece of knowledge for students, practitioners and scholars across the globe.

# Index

www.ingramcontent.com/pod-product-compliance
Lightning Source LLC
Chambersburg PA
CBHW082023190326
41458CB00010B/3253

9 781641 720823